Aquatic weed control

Aquatic weed control

Chris Seagrave
(with a section on weed identification
by Rosie Yeomans)

Fishing News Books Ltd
Farnham · Surrey · England

© Chris Seagrave 1988

British Library CIP Data

Seagrave, C.P.
  Aquatic weed control.
  1. Aquatic weeds. Control measures
  I. Title
  632'.58

ISBN 0-85238-152-2

Published by
Fishing News Books Ltd
1 Long Garden Walk
Farnham, Surrey, England

Typeset by
Mathematical Composition Setters Ltd
Salisbury, Wiltshire

Printed in Great Britain by
Henry Ling Ltd
The Dorset Press, Dorchester

Contents

List of figures 7

Preface 9

How to use this book effectively 11

SECTION A WEED GROWTH AND MANAGEMENT TECHNIQUES 13

1 **Introduction** 13
  Function of aquatic plants in the environment 13
  Classification of aquatic plants 14
  Plant growth 17
  Management strategies 18
2 **Summary of weed control techniques** 20
  Mechanical 20
  Environmental 22
  Biological 24
  Chemical 26
3 **Mechanical weed control** 30
  Handtools 31
  Mechanical devices 34
4 **Environmental weed control** 38
  Control of light 38
  Control of nutrients 42
5 **Biological weed control** 47
  Grass carp 48
  Common carp 55

Silver carp 58
Crayfish 59

6 **Chemical weed control** 62
Role of the Regional Water Authorities 63
Deoxygenation 63
Choice of herbicide 64
Susceptibility of common aquatic weeds to approved herbicides 66
Technical data for recommended herbicides 67
Safety 84
Conclusions 84

7 **Special problems** 85
Algae 85
Rivers 94

SECTION B IDENTIFICATION, CHARACTERISTICS AND CONTROL OF COMMON WEEDS 103
Introduction 104
Keys for identifying problem plants 105
Details of keyed species 111
Emergent plants 112
Floating-leaved plants 122
Submerged-leaved plants 127
Filamentous algae 134

SECTION C 136
Appendices 136
1 Comparative costs 137
2 Addresses of specialised equipment firms 138
3 List of chemicals approved for use in or near water 140
4 Water Authorities in England and Wales 141
5 River Purification Boards and Island Councils in Scotland 145
6 Water pollution control in Northern Ireland 146
7 Other helpful bodies 146
8 Sprayers 147
9 Further reading 149

Index 151

Figures

1 Lake with a well balanced population of aquatic plants *14*
2 The differentiation between the main groups of aquatic plants *15*
3 Shallow pond totally overgrown with floating-leaved plants *16*
4 The range of raw materials required for plant growth *18*
5 Some handtools used for controlling aquatic weed *31*
6 Using a chain scythe in flowing water *33*
7 Using a pole scythe from the bank in deep water *34*
8 Launching a 'Wilder' weed cutting boat *35*
9 Working a weed cutting boat upstream *36*
10 'Bradshaw' weed cutting bucket *36*
11 Maximum outreaches of land-based weed cutting machinery *37*
12 Using a 'Bradshaw' bucket in a small river *37*
13 The shading effect of trees planted on steep banks *40*
14 Too much shade can create 'dead' areas of water *40*
15 The effect of depth on plant growth *43*
16 The three main sources of plant nutrients in lakes and ponds *44*
17 Forming a lake with a bypass *45*
18 Drainable lakes offer much greater management potential *46*
19 Large grass carp *48*
20 Experimental site showing evidence of grazing by grass carp *50*
21 The feeding activity of common carp *55*
22 Mirror carp of approx. 250 g *56*
23 Large silver carp *58*
24 Adult signal crayfish *60*
25 Selective chemical treatment with a knapsack sprayer *63*
26 Large quantities of weed left to die in an enclosed body of water can

have disastrous results *64*
27 Quick guide for identifying common emergent plants *68*
28 Quick guide for identifying common floating-leaved plants *69*
29 Quick guide for identifying common submerged plants *70*
30 Treating emergents with glyphosate 1. tractor spraying *72*
31 Treating emergents with glyphosate 2. boat spraying *72*
32 The result of experimental treatment of lilies with glyphosate *74*
33 Pond surface completely covered with filamentous algae *86*
34 Clearing narrow streams with a hydraulically operated bucket *95*
35 Emergent weed control using tractor-mounted flail mower *95*
36 The shading effects of trees planted on river banks *97*
37 Common weed cutting patterns for a small chalk stream *99*
38 Emergent vegetation is trimmed and trees are carefully pruned to avoid interference with casting *100*
39 The River Test at Leckford with well maintained bankside vegetation *100*
40 Seasonal weed cutting patterns for small rivers *101*
41 Weed cutting in rivers can result in large quantities of weed affecting downstream activities *102*
42 Equipment required for knapsack spraying *147*

Preface

Several years ago, whilst manning an information desk for Sparsholt College at a national 'Country Fair', I was visited by a large number of people asking a similar question: 'I have a lake with a weed problem, what can I do about it?' I explained to each in turn that there were a variety of measures that could be implemented, but in the time available I could do no more than offer some general recommendations on the course of action they should consider.

It was apparent to me that there were many people with water to manage who required information on weed control, but at the time no comprehensive literature was available that I could recommend. It was then that the idea for this handbook was conceived.

I decided to write a book primarily for UK situations, although many of the techniques described have applications for other parts of the world, particularly northern Europe and the USA. I envisage that it should be of value to the owners and managers of a wide range of still waters and rivers which are used for a complete variety of functions (*ie* recreation, ornamental, conservation *etc*).

As the results of weed control can have considerable effects on the biology and appearance of a body of water, this book first asks the question 'What are you trying to achieve?' The answer will have considerable bearing on the choice of control techniques. Once this decision has been made, the reader can start to consider a management strategy.

Section A details a range of techniques currently available for controlling aquatic weed. Each has its merits and drawbacks; there is rarely one which stands out as an automatic choice. The final decision may have to be made after considering a range of factors such as economics, labour availability, site access, time scale or just personal preference.

It will be apparent that particular emphasis has been placed on two techniques from this section. Chemical weed control has been highlighted, as herbicides play a major role in controlling particular groups of weeds. If however their use is mis-timed or their dosage incorrectly calculated, they can inflict terrible damage to the environment which may take many years to rectify.

Grass carp also warrant special reference because of the considerable media attention they have received as 'weed eaters'.

Section B is devoted almost entirely to plant identification, a necessary requirement prior to considering chemical weed control in particular. Plants included in the keys have deliberately been restricted to species which most commonly cause problems. Readers requiring a more comprehensive guide to aquatic plant identification should refer to additional literature.

Section C is composed of a series of useful appendices.

Amongst the numerous people who have helped with the preparation of this book, I would like to give particular thanks to Dr Alan Frake of Wessex Water Authority for providing unlimited amounts of information, to my former student Edward Hopkins for the delightful drawings, and to Rosie Yeomans for the section on weed identification.

I also wish to thank Alan Butterworth of Thames Water Authority, Richard Garnett of Monsanto PLC and J. Bradshaw Ltd for supplying photographic material, and Chris Rogers of MAFF for the processing of black and white prints.

I am indebted to Patrick Haughton, and Alan Grierson for editing the script and to Anne Eade for typing it.

Chris Seagrave
Sparsholt College

How to use this book effectively

When confronted with an aquatic weed problem, fishery managers will find that there are a wide range of control methods available for the weed problems which may be encountered. The eventual choice of technique may depend upon factors such as:

- ease of application;
- site accessibility;
- time scale (for the desired effect);
- labour availability;
- economics;
- personal preference;
- legislative controls.

Occasionally, controlling aquatic weed can be a simple process, with an obvious diagnosis and an equally obvious solution. Conversely, there are many times when weed control problems are anything but straightforward. Even with a fairly comprehensive armoury of techniques available some problems remain difficult to manage.

This book has therefore been designed to provide the information necessary to help the reader decide upon the best possible method of weed control for a given problem. Additional information is also available to provide the manager with the details necessary to use the techniques efficiently.

To use this book effectively, and to ensure that decision making is as easy as possible, the following procedure should be followed:

1 Ask yourself – What am I trying to achieve?/What is a desirable

balance of vegetation in my water in terms of quantity and composition of species?
2 Which weed species are causing problems? – refer to *Section B* (*Identification, characteristics and control of common weeds*).
3 What are the advantages and disadvantages of the various techniques available? – refer to *chapter 2* (*Summary of weed control techniques*).
4 How should the technique be applied? – refer to *detailed text for each method* (find the relevant chapter in *Section A*).

Other considerations may be:

– What are the comparative economics? – refer to *Appendix 1*.
– Where can I buy the tools/equipment *etc* for the job? – refer to *Appendix 2*.

Section A
Weed growth and management techniques
1 Introduction

Function of aquatic plants

Species of aquatic plants are found across a wide range of freshwater environments, from the upland stream to the lowland pond. Each species, whether a marginal or submerged form, will play a major role in the ecology of the system. The essential function of aquatic plants includes:

- the production of oxygen to aerate the water (from photosynthesis);
- the provision of shelter for fish and freshwater invertebrates;
- the consolidation of the river bed and banks;
- the provision of food for aquatic organisms;
- the provision of a spawning medium for many fish;
- marginal plants provide nesting sites and a food source for water fowl;
- the provision of aesthetic appeal to rivers and lakes.

It would be rare, however, to find situations where the growth of aquatic plants maintains a balance which could suit all of man's requirements, particularly when such intensive and varied demands are placed upon the environment. It is common therefore for plant growth to be regarded as excessive, and as 'weed' it is responsible for many problems including:

- flooding;
- silting;
- encroachment;
- blocking of pumps and sluices;
- impeding navigation;
- spoiling recreational activities such as boating and fishing.

It is evident therefore that, as with any garden, the aquatic environment

Fig 1 Lake with a well balanced population of aquatic plants

also requires careful management to produce an attractive and desirable balance of plant life. To achieve this balance it is necessary to understand some fundamental biology of aquatic plants before any control methods are considered.

Classification of aquatic plants (see *Fig 2*)

Emergent plants

Species are found growing in environments ranging from wet ground to shallow water (1–1.5 m in depth). This group contains the erect narrow-leaved plants which include the reeds, rushes, sedges *etc* as well as the broad-leaved varieties such as arrowhead, water-plantain *etc*.

PROBLEMS caused by this group of plants are mainly due to the rapid spreading of the narrow-leaved species leading to encroachment in areas of shallow water. Small ponds once covered in emergents are very difficult to reclaim back to open water without some major restoration work.

FLOATING-LEAVED PLANTS

This group of plants includes two very different forms. Those which are rooted (such as the lily and floating-leaved pondweed) and those which are totally free floating (such as duckweed, and the water fern, *Azolla*).

PROBLEMS specific to this group are related to their potential shading effect. Although most of the rooted species tend to spread comparatively slowly, the free floating plants such as duckweed can completely cover a small pond or narrow ditch in a matter of days. As well as having an unsightly appearance, the dense covering on the surface also restricts the entry of light and oxygen. In severe cases this can produce a deoxygenated and hence sterile environment below (see *Fig 3*).

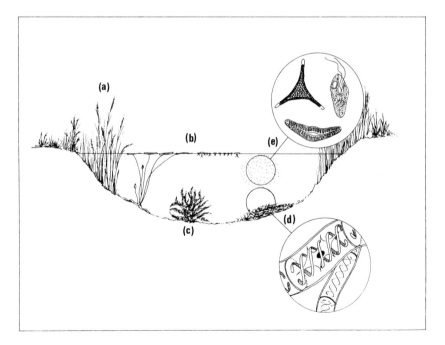

Fig 2 The differentiation between the main groups of aquatic plants found in both still and flowing waters:
a. emergent plants b. floating-leaved plants c. submerged plants d. filamentous algae e. unicellular algae

Fig 3 Shallow pond totally overgrown with floating-leaved plants

SUBMERGED PLANTS

These are characterised by the fact that they are completely reliant on an aquatic existence, as the water column provides total support for the stems. They are particularly important to the environment for the production of oxygen.

PROBLEMS. Like all aquatic plants submerged forms can become excessive in growth and completely choke small shallow waters. Canadian pondweed is typical of such a species. It grows and multiplies prolifically and appears to defy many of the control methods used to combat it.

SINGLE CELLED (UNICELLULAR) ALGAE

This form of algae includes thousands of microscopic species which can be found free floating or attached to rocks, leaves *etc*. Their colour range is commonly green or brown (although many other pigmented forms exist). Like the submerged plants described above, algae is an important oxygen producer within the environment and provides the primary food source for most invertebrate animals.

PROBLEMS. When nutrient levels are high, single celled algae will multiply rapidly and produce a 'pea soup' colour to the water known as a 'bloom', and this may be considered unsightly in certain situations (*eg* in ornamental fish ponds). Other problems associated with algae are that the entire population can die off over a very short period of time when all of the soluble nutrients become used up. The result can be a rapid deoxygenation of the water which can seriously endanger fish stocks.

FILAMENTOUS ALGAE (BLANKET WEED OR COTT)

This form of algae grows in long thread-like filaments which can be attached to rocks *etc* or grow in a free form loosely associated with the bottom silt. Although most species are good oxygenators and provide food and shelter to certain invertebrates, they are generally regarded as an undesirable component of the aquatic flora.

PROBLEMS. When it grows in profusion, filamentous algae produces some very unsightly and troublesome problems. It can rapidly envelop small clear ponds and in larger lakes it can float up, become windblown to one end and choke up very large areas of water. It is regarded as an unattractive group of plants, is generally difficult to completely eradicate and is currently one of the most common weed problems found in still waters.

Plant growth

Once established, the growth rate and size of plants is dependent upon the raw materials which all plants need, namely: sunlight, carbon dioxide, water; plus nitrogen, phosphate, potassium and a range of trace elements (*Fig 4*).

Many plants are perennials which means they die back each winter and then become re-established from stems and rhizomes. Others produce winter buds (known as 'turions') which drop to the bottom in autumn and regrow in the spring.

Under suitable conditions the growth of aquatic plants can be very prolific. For example, the common reedmace (*Typha latifolia*) has been quoted as producing a network of rhizomes three metres (10ft) in diameter

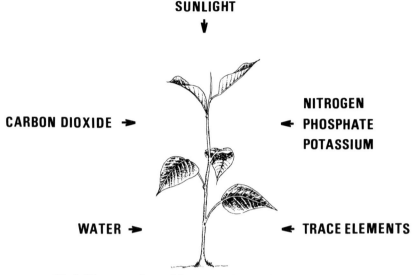

Fig 4 The range of essential raw materials required for plant growth

from one seedling in six months. Equally prolific, the water-crowfoot (*Ranunculus aquaticus*) has been measured as growing into six metre (20 ft) strands in two months during late spring.

Management strategies

It is the opinion of many engineers and biologists in the UK that the quantity of plants in the aquatic environment is increasing due to the very high concentration of soluble nutrients present in water systems (as a result of sewage effluent and fertiliser runoff). The increase in water requirements which encourages greater abstraction, may also serve to exacerbate such problems.

Thus, the increasing problem of weed growth has had a noticeable effect on weed control operations. In certain parts of the country where agriculture is very intensive, standards of aquatic weed control have had to be improved to combat the threat and ensure land drainage continues unimpeded.

Traditionally weeds were removed by hand, but due to the high costs of labour, this technique has become restricted to the small, specialised fisheries. In the important areas for land drainage such as East Anglia and the Somerset drains, machinery and chemicals have provided the only practical solutions. For many situations, however, the complete range of weed control techniques may be considered.

It has already been established that plants are very important to the aquatic environment and for this reason total weed eradication should almost never be considered. The function of weed control should therefore be to reduce plants to acceptable limits without damaging the environment.

Before deciding upon the most appropriate control technique, it will be necessary to highlight the factors which are causing the excessive weed growth, identify the weed species and the problem it is causing and, most importantly, determine the desired end result. Once these features have been considered an efficient weed control programme can be established. It must be appreciated however that plant growth is a continuous process and most management techniques will have to be repeated at least every season (and in some cases two or three times a season) if effective control is to be maintained.

2 Summary of weed control techniques

	Advantages	Disadvantages
Mechanical weed control		
HANDTOOLS		
Digging and pulling	Very effective at removing emergent weed in *small* waters. No regrowth the following season. Very selective.	Very labour intensive and slow to perform, therefore unsuitable for larger waters. All plant material must be removed.
Cutting and hoeing	Can be effective in small waters, particularly for creating suitable conditions for angling. Very selective.	Labour intensive and therefore unsuitable for larger waters. All plant material (particularly submerged plants) must be removed. Regrowth is stimulated by cutting, requiring at least two or three cuts per season.
Raking	Effective for removing windblown weed and creating small 'swims' in coarse fisheries.	Unsuitable for clearing large areas.

	Advantages	Disadvantages
Booms	May be the only suitable technique for floating weeds such as duckweed.	Labour intensive. Immediate regrowth ensures the job is never completed!
MECHANICAL DEVICES		
Weed cutting boats	Very effective at cutting large quantities of submerged and emergent weeds. Suitable for larger waters. Selective. Machines can be hired.	Machines are expensive to buy and even hire if many cuts are required each season. As with hand cutting, regrowth is stimulated. Operators must be skilled and ensure that the enormous quantities of cut weed are removed from the water.
Tractor mounted cutters	Very efficient at cutting emergent weeds along rivers when bankside access is unobstructed. Equipment used is standard agricultural machinery.	Problems of rapid regrowth. Bankside access is frequently unsuitable. Cutting distance restricted by length of hydraulic cutting arm.
Tractor mounted buckets	Very efficient at digging out submerged and emergent vegetation.	Bankside access needs to be suitable for heavy machinery.

	Advantages	Disadvantages
	Large quantities of material is easily pulled out of the water and onto the bank.	Cutting distance is restricted by length of hydraulic cutting arm.
	Worked marginal areas will not be subject to regrowth the following season.	'Dredging' although producing immediate clearance encourages rapid regrowth of submerged species due to viable fragments of weed remaining.
	Cut weed and silt are removed creating instantaneous increase in depth.	
	Conventional digging buckets can be used.	
Specialised (eg Bradshaw buckets)	Efficient at both cutting and removing vegetation above and below water which can be pulled out of the water and onto the bank.	Bradshaw buckets are specialised attachments which would have to be purchased/hired for specific tasks.

Environmental control

CONTROL OF LIGHT

Black plastic sheeting	Inexpensive to purchase and will last for a few seasons.	Unsuitable for larger waters.
	Excellent at killing submerged weeds which should not regrow the following season.	The plastic requires well equipped 'manpower' to install (particularly in deeper water).

	Advantages	**Disadvantages**
	Most suitable for creating 'swims' for angling or specific weed free areas.	May prove unsightly in clear water. May become fouled with anglers hooks *etc.*
Trees	A permanent shading system if sited correctly (particularly for narrow rivers).	Slow to grow and hence produce desired effect.
	Trees provide considerable aesthetic appeal to all freshwater environments.	Deciduous trees will drop leaves into the water which may create additional problems for still waters.
		Trees need to be carefully positioned and pruned to prevent interference with anglers.

CONTROL OF NUTRIENTS

Constructing a bypass	A solution which will *slow down* the onset of silting and weed growth problems.	An expensive and often impractical solution for many lakes.
	The bypass will add considerable management potential to a water.	Existing weed problems will not be affected by this practice.
Draining	Draining can have a drastic effect on stands of submerged weed if the lake bottom is allowed to dry out completely	Many lakes do not dry sufficiently (during a wet summer) to kill off submerged weeds or allow the use of heavy machinery.

	Advantages	Disadvantages
	during a summer. Immediate regrowth will be prevented.	Emergent weeds, once established, will not be affected by periodic draining.
	If lakes become very dry, heavy machinery may be able to work on the lake bottom.	
	Levels may be dropped to allow improved access around the margins.	
	Siltation problems may be reduced.	
	Draining allows other management practices to be enhanced.	
Straw bales *(described under algae control)*	An inexpensive solution to algal problems for some small waters.	Not yet a tried and tested technique for the control of algae.
		Bales must be removed at the end of each season.
Biological weed control		
Grass carp	These fish will graze a wide range of aquatic plants.	Difficult to obtain permission to stock waters.
	They can offer an inexpensive long term solution to weed control in certain	The fish will graze according to stocking density, size of fish and water

	Advantages	Disadvantages
		situations so long as stocks are maintained.	temperature, all of which will vary from year to year.
	They can be successfully used in combination with other techniques.	The fish are expensive to buy as adults. Small fish are vulnerable to predation and will require a couple of seasons before producing a noticeable effect.
	They may selectively graze duckweed when small.	May be deemed an unsuitable mix with trout in some angling waters.
	They make excellent sport for anglers.	
Common carp	In shallow, muddy waters, carp will colour water to such an extent that submerged weed will be rapidly killed off.	High stocking densities are necessary to produce the desired effect.
	Lilies and emergent weeds are not usually affected.	Water retains a muddy colour or develops an almost permanent algal bloom.
	Common carp are not expensive to purchase. Permission to stock is rarely difficult to obtain.	Other fish species do not compete well with high stocking densities of carp.
		Ineffective in deep/hard-bottomed lakes.
Silver carp	These fish will specifically graze single celled algae and *may* have an application in certain situations.	The potential of this fish has not been determined for specifically reducing the concentration of algae in UK waters.

25

	Advantages	Disadvantages
Crayfish	Under suitable conditions high stocking densities of crayfish may serve to reduce the biomass of submerged weed quite considerably. In these situations, a self maintaining population of crayfish would provide an inexpensive and long term control potential.	The fish are not readily available for sale in the UK. Permission will be difficult to obtain for stocking into natural waters. Crayfish are fastidious creatures and will only thrive and reproduce in good water quality. To maintain high stocking densities, the lake may need to have a gravel/rocky bottom to provide suitable shelter.
	Crayfish will provide a valuable cash crop for larger waters and will happily mix with trout.	It may take several years for populations to build up and have a noticeable effect upon the vegetation. Crayfish are predated upon by several bottom feeding fish (particularly eels).

Chemical weed control

General	A small range of chemicals is available for use in water which can control the majority of problem weeds.	A few weeds are particularly resistant. Chemicals are expensive to purchase.

	Advantages	**Disadvantages**
	Chemicals have a lasting effect when plant roots are actively killed.	Over-treatment can result in severe deoxygenation, therefore, to avoid this problem it is often necessary to treat only small areas at a time (submerged weeds).
	The administration of chemicals, whether from a sprayer or as granules, is generally *not* labour intensive.	Dilutions/treatment levels need to be accurately determined to produce the desired effect and to prevent damage to the environment.
	Large areas can be treated in relatively short periods of time.	Nutrient release from dead weed may stimulate algal blooms later in the season.
Glyphosate *(Roundup/Spasor)*	Applied directly to the foliage of emergent and floating weeds therefore very selective.	Is applied late in the season so that the effects will not be realized until the following year.
	Plant roots are killed off preventing regrowth.	The herbicide could be washed away by a sudden rainstorm before it is absorbed by the plants.
		Foliage decomposes *in situ* leaving a silt problem.

	Advantages	Disadvantages
Dichlobenil (Casoron)	Granular formation allows easy broadcasting.	Water depth affects treatment concentration.
	Herbicide is absorbed through the roots allowing localised (selective) treatment of submerged (and some emergent) weeds.	For optimum distribution a mechanical applicator may be necessary.
	Used early in the season and so effect will be obvious soon after treatment.	
Diquat (Reglone/ Midstream)	Diquat is marketed in two forms to suit different situations.	Either form of diquat can be deactivated by muddy water or where silt has been disturbed.
	The gel application allows for very selective treatment. It can also be used effectively in flowing water.	The gel form requires the use of a specialised applicator.
	Diquat will treat some filamentous algae problems.	This herbicide is highly poisonous if swallowed and can be absorbed through the skin.
	Weed can be sprayed some 10 metres from the bank using a gel applicator.	
	Used early in the season therefore results are obvious soon after treatment.	

	Advantages	**Disadvantages**
Terbutryne (Clarosan)	Terbutryne has a wide ranging action against many submerged weed species including many of the filamentous algae.	It is difficult to treat small areas and hence is not very selective.
	Used early in the season, therefore results are obvious soon after treatment.	Considerable care is necessary before treating heavily weeded lakes.
		Algae may require more than one treatment per season.
		The activity of the chemical may be reduced in very 'peaty' situations.

3 Mechanical weed control

Mechanical weed control offers the great advantage of enabling the operator to be very selective in both the amount and the species of weed that is removed at any given time. A variety of techniques can be employed on most waters using a range of either readily available or specialised tools.

The use of handtools would normally be restricted to smaller waters due to the fact that their use is slow and labour intensive. For larger waters, mechanical devices such as weed cutting boats and tractor mounted attachments can provide a rapid and efficient approach to weed control.

Unfortunately, the cutting of weed within a water gives the fishery manager two fundamental problems.

Firstly, the actual process of weed cutting will itself stimulate rapid regrowth, which ensures that at least two major cuts will be needed each season (usually in mid-June and August). An example of rapid regrowth has been noted in the bur-reed *Sparganium erectum*. After a cutting in May plants have been reported as growing some 1 m (3 ft) in only four weeks. Secondly, large quantities of cut weed must quickly be removed from static water before it starts to rot down producing a rapid and severe depletion of dissolved oxygen. In warm weather this can frequently cause immediate and massive fish kills due to asphyxiation. In rivers this is less of a problem as cut weed can be encouraged to float downstream where it can be collected on a suitably placed boom (these can be simply constructed by stringing wire along the surface between two bridge supports).

Handtools (see *Fig 5*)

Digging

This is a thorough method of eradicating rooted plants. If done carefully the work will not require repeating except to prevent encroachment from remaining stands. Although effective it is unfortunately the slowest and most labour intensive method of control. It requires no particular skill but is perhaps best tackled by large groups (when it is not so disheartening!). As for the technique of digging, stands of emergent weeds are best tackled from the inside (*ie* in the deeper water) and worked outside towards the banks. Freed chunks of vegetation can then be pulled into the bank. The rhizome mat should be cut along the edges and cut free from the bottom. Groups should work in a line to ensure complete coverage.

Once plants have been removed an immediate increase in depth of some 150 mm (6 in) can be achieved. It is difficult however to dig effectively in water greater than 0.6 m (2 ft) in depth.

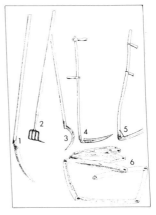

Fig 5 Some handtools used for controlling aquatic weeds:
1. pole scythe
2. weed rake
3. slasher
4. Turk scythe
5. hand scythe
6. chain scythe (or 'links')

NB. The best tools for the job are conventional garden spades and forks, although specially lengthened tools may offer greater versatility.

Cutting and Hoeing

Slashers

Slashers are made with either a curved or a straight blade attached to a 1.5 m (4–5 ft) wooden handle. Both types are useful for cutting plants down to the roots in shallow water. They are also useful for the control of emergent weeds and other bankside vegetation.

Scythes

The effective use of a scythe is a skill which can only be developed with practice. For 'aquatic' work the hollow metal shaft ('Turk') scythes are lighter and easier to handle than the large wooden types. Like the slasher the scythe can only be used where wading is possible. It will also need repeated sharpening to maintain efficiency.

Chain scythes

Chain scythes are sometimes referred to as 'links' as they are constructed of a series of loosely-linked blades which enable them to flex. A set of links consists of a variable number of blades connected to a rope at either end. In use, the operators pull on the ropes and work the scythe in a saw-like motion as they walk forward along the banks (*Fig 6*). Consequently, this tool will only be effective in rivers and lakes where the bank width does not exceed say 50 metres. By working upstream all weed can be cut by slowly working the blades along the bed. The slower the action, the deeper the cut. Downstream cutting offers the operators greater control and allows for patches to be left.

Pole scythes

River keepers on the larger chalk streams (like the River Test) often construct a pole scythe for trimming submerged weed in deep water close to the bank. These instruments consist of a conventional scythe blade bolted to a long aluminium handle some 5 m (16 ft) long. The blade is worked through the weed in a 'jerking' motion back towards the bank (*Fig 7*).

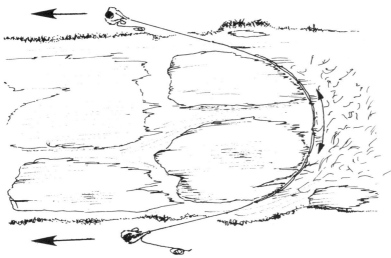

Fig 6 Using a chain scythe in flowing water (working upstream)

Hoes

Richard Seymour, in his delightful book on river management, describes a tailor-made hoe (much stronger than the garden variety) which he designed for the chopping out of roots and shaping weedbeds. It is a quick and easy job in shallow water (particularly flowing). Unfortunately, cutting without removing roots encourages faster growth and may need to be repeated up to five times a year.

Rakes

Conventional strong garden rakes, tied to lengths of rope can be used for clearing small areas of submerged plants and for removing floating weed and blanket weed. It is not a very efficient tool unless the weed has been allowed to concentrate into a corner by the wind (*NB*. Strong winds can be a useful ally to a still water manager during weed cutting exercises). Regrettably, algae and duckweed multiply so quickly in fertile water that raking only produces a short term effect.

Booms

Free floating weed (duckweed in particular) can prove to be extremely

Fig 7 Using a pole scythe from the bank in deep water

difficult to control and may require a specialised gathering device. Shallow drafted seine nets or even a narrow strip of chicken wire can be used (if sufficient hands are available for pulling) or, for a one man operation, a collecting collar can be constructed. This collar (designed but not patented by a local carp farmer!) consists simply of a few pieces of wood, nailed into a 'U' shape. This collar is then pushed around in front of the operator. It is an effective tool because the weed does not sink down as it is approached. The resultant 'catch' can be deposited into a corner for later removal.

PULLING

In the autumn, emergent plants (particularly reedmace—*Typha* species) can be pulled up by hand. This action should also result in the removal of the tuber which will prevent regrowth the following season. Small areas of emergents can be conveniently controlled by this method.

Mechanical devices

Weed cutting by hand is always hard work, particularly as the cut weed often has to be removed from the water at the end of the day. Long stretches of river and lakes larger than say one acre may therefore require a more mechanised approach.

WEED CUTTING BOATS

Specialised weed cutting boats are available for purchase from one or two specialised engineering firms. They are shallow-drafted vessels driven by a

Fig 8 Launching a 'Wilder' weed cutting boat

paddle wheel or screw (a propeller would quickly become clogged with the cut weed). Most models have reciprocating cutting blades, like those of a hedge trimmer on a U or ⊥ shaped beam. This beam can be adjusted to cut at any depth down to a maximum of some 2 m ($6\frac{1}{2}$ ft). A good weed cutter should be light and easily handled by one man. One such boat available in the UK comes with its own special trailer, which although expensive, facilitates easy removal of the boat to and from the water and enables it to be transported to different sites (*Fig 8*).

Many weedcutting boats are owned and employed by Water Authorities, particularly in the Fenlands for clearing drainage channels. In other parts of the country (particularly Hampshire) some river keepers own new/secondhand boats and hire themselves out on contract for regular weed cutting programmes particularly in chalk streams and in trout fisheries (*Fig 9*). Their hire has proved to be an economical proposition for local fishery managers.

BUCKETS

As well as boat-mounted cutters there are other examples of specialised equipment designed specifically for weed control (*Fig 10*). Mechanical

Fig 9 Working a weed cutting boat upstream

Fig 10 'Bradshaw' weed cutting bucket

excavators may be equipped with a weed cutting or roding bucket which cuts and removes plants at the same time as dredging a channel. This will work effectively in narrow channels where the arm of the machine can reach (see *Fig 11*). The bucket consists of a series of curved bars connected by a horizontal bar at the top. Another bar at the bottom has a reciprocating cutter blade to form the lower edge. As the operator moves the bucket it cuts and collects weed and allows the water to drain out through the bars (small versions of these buckets are also available for backactors on conventional tractors).

WINCHES

Occasionally, emergent plants such as reedmace (*Typha*) and floating sweet-grass (*Glyceria*) form floating 'rafts' of vegetation. These can be drawn to the bank using a winch after the raft has been cut into manageable pieces.

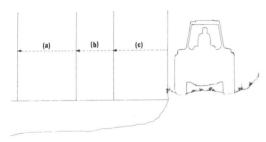

Fig 11 Maximum outreaches of land-based weed cutting machinery:
a. Weed cutting bucket on dragline (18 m)
b. Weed cutting bucket on excavator (11 m)
c. Tractor mounted flail mower (6.5 m)

Fig 12 Using a 'Bradshaw' bucket in a small river

4 Environmental weed control

Plants need three basic requirements to grow. They are LIGHT, NUTRIENTS and a suitable TEMPERATURE. By reducing or eliminating one or more of these essential requirements then plant growth can be slowed down or even prevented.

Control of light

The exclusion of light by some form of shading does in fact offer some quite practical possibilities for weed control. It can be approached in a variety of ways:

BLACK PLASTIC SHEETING

Black plastic (polythene) sheeting is occasionally used in still waters to create weed-free areas. Strips of 1500 gauge polythene can be used singly or heat sealed together to produce a large area of light-excluding material. Regular small holes should be punched in (rows 0.5 m ($1\frac{1}{2}$ ft) apart, holes every 0.3 m (1 ft) to allow gases produced by the underlying mud to escape. The sheets are rolled out in the selected area and weighed down with old scrap iron, bricks etc. The operation may require the use of a diver in a wet suit and mask particularly if the water is over say 1.5 m (4–5 ft) deep.

The plastic can be left in place 'permanently' to produce weed-free areas for the entire season, or it can be moved every three or four weeks. This latter method will kill off areas of weed, but then may become recolonised towards the end of the season.

The advantage of using black plastic is that it is inexpensive to purchase and once installed may be effective for several years. The drawbacks are that even when using punched holes, some bulging can occur particularly when the pond bottom is excessively muddy and large volumes of gas are produced. The plastic can however be easily 'prodded' down when necessary. This problem may be encountered less on ponds with cleaner bottoms. Plastic also tends to look unsightly although it should be less readily noticed in deeper and more coloured water.

In his book on the management of angling waters, Alex Behrendt describes the use of black plastic at his 'Two Lakes' trout fishery. Two men each take a corner of the sheet (16 m × 8 m) and pull away from each other to tension it. The sheet can then easily be slid under the surface of the water. Behrendt found that most weed is killed in four weeks; the sheets are then moved to a new location. Plants will reappear the following year unless the sheets are left for at least three months. This longer duration should kill the roots and create weed-free areas for a couple of years.

TREES

As well as providing aesthetic appeal to most environments trees can, in certain situations, provide enough shading to produce some control of weed growth. However, factors such as positioning, choice of species and management of the trees must be carefully considered before planting begins.

In rivers and small narrow lakes, mature trees can provide enough shading to produce complete weed-free areas. They also keep water temperatures lower. Where the topography permits, a combination of a steep bank and tall trees can produce a considerable area of shade along a bank (*Fig 13*).

The correct siting of trees is therefore particularly important, and depends upon factors such as tree species, orientation of bank, size of river etc.

Common species

Alder (*Almis glutinosa*)
This is probably the most common species found lining rivers and lakes. It is fast growing and has a root system which binds the river banks well. It does, however, rapidly become top heavy and should be pollarded every few years (before reaching 10 m (33 ft) in height). Unmanaged trees can

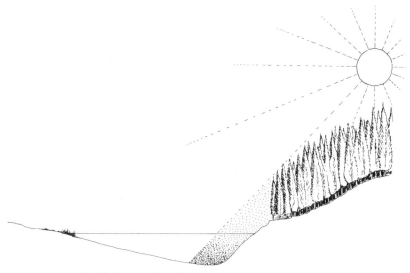

Fig 13 The shading effect of trees planted on steep banks

Fig 14 Too much shade can create 'dead' areas of water

easily be blown down destroying much of the bank they had previously protected.

Crack willow (*Salix fragilis*)
Another good shade producer for small streams but again must be pruned every few years to prevent excess growth. These shallow-rooted trees are easily blown over.

Weeping willow (*Salix alba babylonica*)
A fast-growing tree with good shading properties. It has a much stronger root system than the crack willow.

Suitable trees for shading

Close to the water:
Alder
Elder
Crack willow   All require regular surgery to produce best effects
Weeping willow
Goat willow

Further back on the bank:
Maple	– plant in groups of four
Hawthorn	– slow growing
Scots pine	– good height (relatively poor shade)
Poplar	– plant in groups
Larch	– plant in groups

Unsuitable for river banks:
Beech	– leaves slow to rot, prefers dryer ground
Oak	– leaves slow to rot, prefers dryer ground
Lime	– prefers dryer ground
Elm	– prone to diseases
Hornbeam	– leaves slow to rot, prefers dryer ground
Plane	– prefers dryer ground
Chestnut	– prefers dryer ground
Ash	– little shade
Birch	– grows spindly if too wet

There are however some important disadvantages to having trees too near water particularly:

- Water loss – trees such as alder utilise vast quantities of water for transpiration. This could be significant to small bodies of water.
- Leaf fall – may present problems in small shallow lakes where water quality may be affected as well as providing a contributory factor in increasing the rate of silting.

It must also be remembered that trees should never be planted on dams as serious leaks could occur as a result of root damage.

Control of depth

A key physical characteristic of still waters which can have a marked effect upon plant growth is the depth.

If a lake is deep enough plant growth will be restricted for the following reasons:

- Emergent plants such as reeds and rushes have definable maximum depths from which they can grow, and this is usually around 1–2 m (3–6½ ft). This restriction can be frequently observed in old gravel pits where emergent plants form only the narrowest of fringes around the margins. This is because the depth falls away so sharply at the edge.
- Submerged plants will only grow at depths where light can reach. The deeper the water the less light can penetrate to the lake floor. Unfortunately, the depth at which plants will *not* grow cannot be clearly defined as it varies from water to water and throughout a season due to natural variations in water colour, turbidity, unicellular algae content, fish activity *etc*.

However, to increase the depth of an existing body of water in an attempt to control plant growth is rarely a practical proposition, but when new lakes are in the planning stage it would be worth considering the depth and contours of the lake floor to help reduce weed problems for the future. *Fig 15* demonstrates how the provision of both shallow and deep areas will keep the eventual plant growth naturally in check.

It should be noted however that in many spring fed lakes, or where water clarity is particularly high, light can penetrate to considerable depths [over 6 m (20 ft) would not be uncommon].

Control of nutrients

Plants require a complete range of nutrients for growth and reproduction.

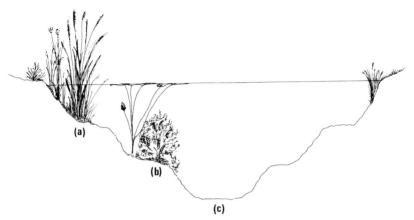

Fig 15 The effect of depth on plant growth:
   a. Shallow shelf to support emergents (approx. $\frac{1}{2}$–1 m)
   b. Deeper shelf for submerged and floating-leaved plants ($1\frac{1}{2}$–3 m – emergents cannot invade this depth)
   c. Deep area to prevent light penetrating to lake floor (3 m +)

These nutrients are found in varied quantities in different bodies of water at different times of the year; they fall into three main categories:

- primary nutrients (required in large quantities) – Nitrogen (N), Phosphorus (P), Potassium (K).
- secondary nutrients (required in small quantities) – *eg* Calcium (Ca), Sulphur (S), Magnesium (Mg) *etc.*
- trace elements (required in minute quantities) – *eg* Copper (Cu), Zinc (Zn), Iron (Fe), Manganese (Mn) *etc.*

When one or more nutrient is scarce in an environment, then aquatic plant growth may be minimal (*eg* in upland rivers and lakes). However, when nutrients are present in abundance, then plant growth may be excessive (as long as there are no other limiting factors). This is often the case involving lowland waters which receive a 'runoff' of nutrient-rich water (particularly containing high concentrations of nitrates and phosphates) such as from land drainage, sewage or farm effluent. This continual renewal of nutrients ensures plants can grow unhindered through the season.

If it is understood where nutrients originate, enclosed waters can often be improved with some comprehensive management. *Fig 16* indicates

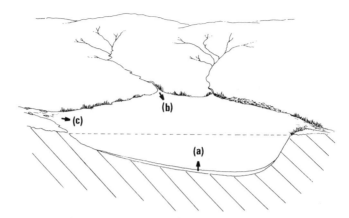

Fig 16 The three main sources of plant nutrients in lakes and ponds:
a. bottom silt b. run-off from surrounding land c. inlet water

three main sources of input to a lake.

In severe cases of 'weeding up' the following (fairly drastic) control measures should be considered.

FORMATION OF A BYPASS

There are many good fishery management reasons why lakes should be constructed with a controllable inlet and a bypass channel, particularly where lakes are fed by a small stream. With this arrangement the inlet should be stopped off so that all further incoming water can be diverted around the side, to join the old stream bed below the lake (*Fig 17*).

If incoming water cannot be diverted, the lake will receive a continuous input from the stream which has several disadvantages. Rainfall will result in large quantities of silt being washed into the lake. As well as colouring the water and reducing depth, this process adds to the level of available plant nutrients within the environment.

Although a new bypass will not in itself reduce weed problems it should help in 'slowing down' the problem. It will, however, be an essential requirement if a complete draining is required (see next section).

At such times of low water, the inlet can of course, be opened up to replenish levels. The inlet will also be useful in providing a short 'flush' to the lake after say chemical treatment, or at times of algal bloom. In both cases, this will serve to dilute the high concentrations of soluble nutrients present.

Draining

The ability to drain a lake (*ie* by use of a sluice or monk), offers a wide range of management possibilities which can be applied to weed control. Complete draining to allow the bottom to dry out for a period of time, particularly over summer, can serve to kill many aquatic weeds, especially the submerged varieties. On occasions, it may also offer the possibility of utilising heavy machinery such as rotovators to break up more resilient plant roots, or even bulldozers to push out excess silt (*Fig 18*). This effectively gives the lake manager the opportunity to start afresh, which may be essential when lakes have become neglected over long periods of time.

It must be considered however, that with a typical UK climate badly silted ponds may never dry out sufficiently to offer this possibility unless the summer is particularly long, hot and dry!

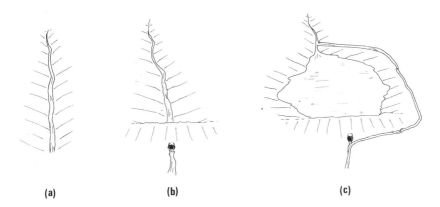

Fig 17 Forming a lake with a bypass:
    a. stream running through a shallow valley
    b. the valley is dammed to create a lake
    c. the channel is created to divert the flow around the lake. It rejoins the original stream bed behind the dam

Fig 18 Drainable lakes offer much greater management potential

5 Biological weed control

Biological control involves the use of a living organism to control an animal or plant pest. There are many successful examples of this type of control in use today. These range from the simple domestic use of pigs and goats to clear brambles from areas of land, to the enormous government sponsored programmes which were set up to control problems on a national scale (*eg* the introduction of the Argentinian moth borer to destroy the prickly pear cactus in Australia).

Unfortunately the success of biological control can only be determined by trial and error particularly when a foreign species of animal is being introduced into another country. Some spectacular mistakes have been made in the past where problems have been exacerbated rather than reduced by the controlling agent itself.

Biological control is, however, a particularly attractive concept in certain situations as:

- it does not require the use of toxic chemicals;
- it may continue to be effective over very long periods of time (particularly if the agent can reproduce successfully);
- it requires little, if any physical labour.

Regarding aquatic weed, a degree of biological control can be achieved using a variety of animals, and in particular certain species of fish. This chapter has therefore been devoted to the biological control of aquatic weeds by the following animals:

- grass carp;
- common carp;
- silver carp;
- crayfish.

Fig 19 Large grass carp (approx. 9 kg). A fish this size can consume considerable quantities of plant material in a day

Grass carp

The grass carp (*Ctenopharyngodon idella* Val) is also referred to as the 'Chinese grass carp' and the 'White Amur' (in the USA). These names reflect the origin of the fish which is the large eastern flowing rivers of northern China; in particular the River Amur.

This species of fish belongs to the 'cyprinid' family and is generally regarded as a true herbivore. With this in mind, it was introduced into the USA and most European countries during the 1960s specifically for weed control purposes.

Its appearance is superficially like that of the chub (*Leuciscus cephalus* L.), a fish for which it is often mistaken. However, unlike the chub, the grass carp has an almost exclusive diet of aquatic plants and grows much larger (up to some 15 kg, 33 lb).

Although sub-tropical in origin, it can survive a wide range of water temperatures (0–35°C), and withstand low concentrations of dissolved oxygen (<1 ppm). It is a fast growing animal and its gut is short (approximately $2\frac{1}{2}$ times body length), which is not generally characteristic of a true herbivore. Neither does it possess cellulases, or a facility for bacterial digestion. The fish does however, possess a well developed set of pharyngeal teeth which facilitate the crushing and shredding of plant material. However, even with these teeth, the fish only manages to digest approximately 50% of the food ingested.

As with all fish species, the feeding stimulus is primarily determined by water temperature. Grass carp do not actively feed much below 16°C, but at their optimum feeding temperature of 25°C, they can consume their own weight of plant material in one day.

GRASS CARP IN THE UK

Following some successful introductions in Europe, about 20 years ago, the first stocks of grass carp were brought into the UK for use in weed control trials. Within a short time of their arrival there was a considerable amount of speculation concerning their potential as a weed control agent.

The press immediately seized upon the idea that grass carp might provide the much-vaunted solution for fishery managers with weed control problems, and the availability of the fish was eagerly awaited. However, a few years later the grass carp was still not readily available for stocking.

Reasons for this delay were probably in response to an earlier experiment with an exotic fish species which had been subject to considerable criticism. A decade before, the decision was made by a water board to introduce the 'zander' (*Stizostedion lucioperca* L.) into UK waters. The stocking of the zander into one particular river system resulted in a rapid population explosion of this carnivorous fish. Within a few years, numbers of other fish species were seen to decline, and the zander was immediately blamed by the angling fraternity. Many people saw this introduction as an ecological disaster parallel almost to that of the rabbit in Australia!

The Ministry of Agriculture Fisheries and Food (MAFF) who were conducting these early grass carp trials, understandably did not want to make a similar mistake. With no individual prepared to authorise the release of the grass carp, more and more trials were conducted to assess the fish as a weed-controlling agent in UK situations, but more importantly, to assess its likely impact upon the environment.

Now, some twenty years after that initial importation of fish, Water Authorities have finally allayed many of their fears and have started to give consent (albeit strictly controlled) to stock certain UK waters.

POSSIBILITIES OF NATURAL SPAWNING

The introduction of any new species into a country, must of course take into account the possibility of serious ecological damage by naturally breeding populations. Wholesale destruction of aquatic plants could be disastrous and obviously unacceptable. Results of introductions in the USA and other European countries however, have demonstrated that natural reproduction does *not* readily occur.

In its natural environment, the grass carp seems to demonstrate the need for some very specific conditions, namely a temperature of 21°C, a current speed of 0.5–2.4 metres/second, and a sudden rise in water levels. This ensures that the eggs are kept afloat for a long enough period of time (some two days) for the fry to hatch out.

This combination of conditions does not occur anywhere within the UK. The fish will not spawn in still waters, and our rivers are not warm enough or long enough (some 200 kilometres would be necessary) to ensure successful hatching. Although it would be very difficult to prove that grass carp would definitely not spawn, the chances are very remote indeed.

STOCKING RATES – RESULTS OF TRIALS

During the last 20 years, a variety of trials and studies have been conducted in a variety of climates around the world to assess the weed controlling efficiency of grass carp.

Fig 20 Experimental site showing evidence of grazing by grass carp

Most countries reported some success in controlling aquatic weed. Results from typically temperate climates (*eg* UK and parts of Europe), have shown that weed control efficiency may also be high, costs are comparatively low, and no severe side effects (in terms of damage to the ecosystem) have been observed.

Stocking densities must of course vary between climatic areas as the feeding rate and duration of the season are so critically controlled by temperatures.

UK trials have led to the following observations and suggestions concerning stocking and control:

- Grass carp, being naturally a sub-tropical fish will only graze effectively when the water temperature has reached 16°C. Optimum feeding conditions (which occur when water temperatures are maintained between 20–28°C/68–82°F) can therefore only occur for short periods of time within a typical UK season, and may vary between parts of the country.
- The grazing power of the fish depends on the stocking density. The stocking density (or biomass) of fish is measured by the number or weight for a given area; usually stated in kg/hectare (which relates very closely to lbs/acre, and for convenience may be regarded as the same).

    There is no hard and fast stocking density which can be quoted to guarantee a particular weed control, as there are too many variable factors which influence the result (in particular, temperature, weed type, water depth, size of fish, and time scale for desired control *etc*).

    As a guideline, the following figures may be acceptable:

(1) 500 kg/ha of 'large' fish (say 500 gram or 1lb in weight) is a high stocking density which should provide fairly rapid results. This approach will be expensive and could result in an excessive degree of control.

(2) 200 kg/ha of 'smaller' fish (say 100 g or 4 oz in weight) will build up over two or three years to provide an increased biomass which should ultimately give effective control. This method is far cheaper, but slower in effect.

- It should be stated that the excretions of grass carp lead to continual enrichment of the aquatic environment with nutrients. In some cases, this may lead to the production of algal blooms. These blooms may create problems in certain situations (for more information see *chapter 7*).

It is obvious that biomass will have to be maintained to ensure adequate control is available over a period of years. This may not always be readily achieved for a variety of reasons. The grass carp will inevitably grow (at times quite rapidly), resulting in the biomass increasing throughout the season. If this biomass becomes too large, and conditions are optimal for feeding, overgrazing will occur, resulting in the complete removal of all aquatic vegetation. (Even the marginal reeds, rushes and sedges may be attacked!)

In contrast, excessive mortalities within the grass carp population will seriously reduce the biomass, resulting in inadequate weed control. In temperate climates, it is common to find a natural mortality of many fish species during the winter and early spring: grass carp are also susceptible over this period.

Predation by pike is also a serious problem particularly amongst smaller grass carp, which appear to fall easy prey to this native carnivore. With grass carp unable to spawn in still waters, natural recruitment cannot possibly occur and the biomass will ultimately decline. It is apparent that optimum weed control can best be achieved if the population can be assessed and regulated over the course of time. The ideal method to assess fish stocks is in a *drainable* lake, as these environments always permit far greater management control. As well as removing grass carp (if required) pike can be easily culled to protect stocks. Undrainable lakes unfortunately are more difficult to manage and if they contain a large biomass, it will be difficult to assess accurately, and effective fish removal may be almost impossible.

From results obtained in the UK, Fowler (1982) has drawn up a 'preferred grazing' list. In general, 'soft vegetation is preferable to the fibrous plants unless stocking densities become excessively high when almost any vegetation will be consumed by larger fish.

Plants are listed in order of preference.

Readily eaten
Duckweed (*Lemna* spp)
Stonewort (*Chara* spp)
Canadian pondweed (*Elodea* spp)
Starwort (*Callitriche* spp)
Small-leaved pondweeds (*Potamogeton* spp)

Less readily eaten
Hornwort (*Ceratophyllum* spp)

Milfoils (*Myriophyllum* spp)
Water moss (*Fontinalis* spp)
Mare's-tail (*Hippuris* spp)
Water buttercup (*Ranunculus* spp)
Large-leaved pondweeds (*Potamogeton* spp)
Filamentous algae (*Enteromorpha* and *Cladophora*)

Eaten only by large fish in overstocked situations
Filamentous algae (*Vaucheria* and *Spirogyra*)
Water lilies (*Nuphar, Nymphaea* spp)
Emergent reeds/sedges (*Typha, Carex, Phragmites etc*)

ANGLING POTENTIAL

Although grass carp have been regarded primarily for weed control purposes, they also offer exciting possibilities for exploitation as a sport fish. Under suitable conditions it has been shown that these fish can grow very rapidly. For example, fish stocked at 100 g (3–4 oz) should only take a couple of seasons before their size reaches some 700 g ($1\frac{1}{2}$ lb). The fish can grow to some 9–13 kg (20–30 lb) thus it may not take very long before waters contain a number of quite large specimens.

Data available from the limited number of waters which support a population of grass carp has confirmed what has already been found in the USA and other countries; that the fish are very popular with anglers. They have been reported to take common baits such as bread, maggots *etc.* but unlike most of the indigenous fish of the UK grass carp will often jump well clear of the water providing a most spectacular fight on light tackle. As its optimum feeding range suggests, the best angling will be found in the warmer summer months rather than the winter.

It is also worth noting that in a mixed fishery, the increased productivity resulting from a population of grass carp may indirectly improve the growth rates of other fish species.

Once the stocking of this fish becomes more widespread, many fisheries managers are convinced that it will not be long before the fish really catches the imagination of the 'coarse' angling fraternity. One particular group ('specimen anglers') are willing to devote a considerable amount of time and money in pursuit of large fish. Entrepreneurial fisheries containing a good stock of grass carp may find a considerable amount of new and lucrative interest shown by this type of angler, particularly now that a serious rod caught record has been established (standing at 14 lb).

Legislation concerning the stocking of grass carp in the UK

The movement and hence stocking of all fish in the UK, is controlled by the Salmon and Freshwater Fisheries Act, 1975 (section 30), and enforced by the Regional Water Authority (RWA). The introduction of grass carp into any water therefore requires a stocking consent.

Until recently, this consent to stock grass carp, was almost impossible to obtain by private individuals, as research trials were still being conducted. Today, some Water Authorities seem satisfied that the fish is 'safe' to introduce into enclosed waters and permission to stock is therefore more readily available. In fact, fishery staff within the Water Authorities are happy to provide fishery management advice regarding the stocking of grass carp in individual situations.

To satisfy all legal requirements, special attention also has to be paid to the Wildlife and Countryside Act; section 14 refers to the introduction of exotic animals into the 'wild'. As the grass carp is specifically mentioned in this category, its stocking necessitates the purchase of a special licence which is issued for an individual water.

To be considered exempt from this licence, the applicant has to demonstrate that the fish are *not* being introduced into the 'wild'. As the wild is not clearly defined within the Act, it leads to some considerable confusion regarding its interpretation. For example, a garden pond would be regarded as exempt, but what about an enclosed lake? Until clarification of this situation is established, there will inevitably be confusion regarding the necessity of a licence. Applicants will just have to appreciate that each case will be taken on its merit, and results may not always be consistent! The Nature Conservancy Council (NCC), which advises the UK Ministry (MAFF) on granting approval, will make the final decision.

Stock availability

To date, grass carp are unfortunately only available from a small number of sources. As the importation of these fish is strictly controlled, the only legal sources within the UK are from a handful of commercial fish farms. Consequently, the price of fish is rather high as the demand currently far exceeds the supply.

Small numbers of fish however, may be available from local Water Authorities; and as their permission is a necessary requirement prior to stocking, it could be prudent to establish whether they have some available.

However, with the increasing numbers of cyprinid fish farmers in the UK, availability will very soon improve.

(See *Appendix 2* for list of suppliers).

CONCLUSIONS

Although trials have been conducted now for some years, there is still a considerable lack of basic information regarding the potential of grass carp in the UK. Only time and experience will determine whether they will prove to be a viable and economic proposition for controlling aquatic weeds.

Common carp

Fish which are considered under the name 'common carp' also include the mirror, leather and wild carp which all represent varieties of *Cyprinus carpio*.

Fig 21 The feeding activity of common carp

This species of fish can effectively control aquatic weed in certain lakes primarily because they are active bottom feeders and are constantly foraging deep into the mud in search of food. As they are, by nature both large and strong fish, this feeding behaviour results in aquatic plants being continually uprooted, which proves to be fatal to many species. Carp will also directly consume the newly sprouting shoots of many plants as a normal part of their omnivorous diet.

As a result of their feeding behaviour, carp also contribute to weed control by indirect means. The process of feeding, which can be very active in warm weather results in the disturbance of large volumes of mud/silt which become suspended, and hence cloud the water. This process leads to a reduction of light penetration through the water which considerably reduces the growth of submerged plants. Another consequence of this activity is that increased amounts of chemical nutrients (such as phosphates and nitrates) are released from the bottom mud and dissolve in the water. All plants will readily utilise these nutrients but none more so than the unicellular algae. These microscopic organisms present

Fig 22 Mirror carp of approx. 250 g, the minimum required size for effective weed control

in all bodies of water can multiply very rapidly to produce a green colouration to the water. This is termed an 'algal bloom' and in extreme cases it may become so intense as to prevent light penetrating more than a few inches below the surface. This further shading will completely kill off any submerged plants and prevent their recolonisation so long as the bloom persists.

To achieve this level of weed control it is essential that carp are introduced at very high stocking densities. Figures accepted for weed control are usually 2,500 fish/ha at 250 g + (approximates to 1,000 fish/acre at 8 oz = 500 lb/acre).

This level of stocking may well be considered expensive and therefore the most economical way to achieve this biomass would be to stock the desired number of fish in spring at one year old (when they weigh 15 g/½ oz). After one summer's growth in a reasonably fertile lake they should grow to 170–260 g/6–8 oz, and by the following season (*ie* one year after stocking) the desired biomass should be achieved.

Carp are very hardy fish and enjoy a long life (15 years +). If the population proves to be self sustaining weed control should continue to be effective for many years.

In some waters, very high stocking densities of carp have also been known to check the encroachment of marginal plants such as the reeds and sedges *etc*. The continual grazing at the base of these plants by large fish appears to uproot tubers and roots. Small 'rafts' of these emergents then break free and eventually become destroyed. Over the years a gradual reduction in the area of emergents may be observed; eventually only a narrow margin close to the bank may remain. It is worth noting that lilies do not seem to be affected in this way.

As might be surmised, the most effective results obtained by using carp for weed control are in muddy ponds or lakes. Gravel pits cannot be grazed so effectively as the fish cannot penetrate into the hard bottom.

As with most fish species the temperature of the water will determine the feeding activity. Carp are most active in warm summer temperatures (with an optimum of approximately 25°C/77°F) and therefore shallow lakes (say 1–2 m (3–6 ft) deep) warm up quickly and may prove to be most suitable.

CONCLUSIONS

If the correct stocking densities are maintained, common carp should prove to be an effective, cheap, and long-term answer to submerged weed control in small, shallow lakes and ponds. Achieving this level of control may, however, result in some characteristic changes to the environment.

The most common feature is the distinctive green colouration of the water produced by blooms of unicellular algae. This appearance may not always be considered desirable, as fish cannot be readily seen as in clear water. However, some individuals find 'green' water most attractive, particularly where beds of lily pads are present.

Furthermore, intense blooms can occasionally create water quality problems, as unicellular algae have an unusual characteristic in that the entire population can suddenly die off over a two or three day period. This bloom 'collapse' usually causes a rapid reduction in the dissolved oxygen level of the water which in some cases can be quite drastic. Mostly, cyprinid fish such as carp, roach, goldfish *etc* would not be unduly affected by these short-term low oxygen levels, but they could prove fatal to fish with high oxygen requirements such as trout. (For more information regarding algal blooms see *chapter 7*).

Silver carp (*Hypophthalmicthys molitrix*)

Like the grass carp, the silver carp is a fish native to the sub-tropical regions of Asia, particularly China. It is most unusual in the fact that it

Fig 23 A large silver carp. An interesting 'exotic' fish, but unlikely to effectively control algae

feeds exclusively upon unicellular algae, which it strains from water by means of specially adapted gill rakers.

As the silver carp has such a specific food preference, fishery scientists quite naturally looked at the fish as a potential tool for controlling unicellular algae. Results however were generally disappointing. Although the fish feeds and grows well, the algae it is meant to control appears to reproduce faster than it is consumed.

CONCLUSIONS

Although there has been considerable interest in this fish recently in the UK, the conclusions drawn from an albeit limited amount of research is that within the constraints of a European climate, the silver carp cannot offer any significant control for unicellular algae problems in lakes and ponds. There may however be some potential for 'special' situations (perhaps even in garden ponds) but as yet there is little scientific evidence to support this.

Crayfish

The potential for using crayfish specifically as weed controlling agents has never been comprehensively assessed. Only a few scientific papers have even touched upon the subject. It is, however, a fact that most crayfish species will eat aquatic vegetation as part of their omnivorous diet, and where the stocking density of these animals is particularly high, some control over weed growth may be achieved. This fact emerged from a Swedish report which noted that submerged weed growth in one particular lake suddenly became very prolific the year following a complete crayfish mortality. The crayfish have still not recolonised the lake and the weed is now a major problem.

Current evidence suggests that where a level of weed control is achieved by crayfish it is due to the feeding habits of the animals which is to cut large quantities of plant stems near the lake floor, rather than by actual consumption.

Crayfish are quite fastidious animals and have clearly defined habitat preferences. They need reasonably hard water (with a calcium carbonate level in excess of 100 ppm), which will warm to above 15°C (59°F) during the summer to stimulate crayfish growth. They prefer lakes with fairly clean lake floors (such as gravel pits) rather than those with deep layers of mud.

Fig 24 An adult signal crayfish. Such animals in high density have been recorded as controlling weed in Sweden

The main species of crayfish grown and available in the UK and Europe are as follows:

- native British crayfish – (*Austropotamobius pallipes*)
- European (or noble) crayfish – (*Astacus astacus*)
- signal crayfish – (*Pacifastacus leniusculus*)

Of these three species listed the first two are indigenous to Europe whereas the signal crayfish is a recent import from the USA.

The signal crayfish* grows much larger and faster than the European species and is also resistant to a fatal disease called 'crayfish plague' which has caused considerable mortalities amongst the native stocks.

Crayfish have become the subject of considerable interest during recent

---

*It should be noted that signal crayfish are regarded as an exotic species and may be subject to control under the Wildlife and Countryside Act in the UK as specified in the 'grass carp' section.

years as many individuals are attempting to 'farm' these animals in lakes and gravel pits. Mature animals are sold to restaurants and wholesalers and can represent a quite lucrative crop. To have any control over weeds however, stocking densities should be kept as high as possible. Heavy cropping of the crayfish population will therefore not be conducive to effective weed control.

A list of crayfish suppliers is given in *Appendix 2*. However, it is well worth seeking advice from a reputable supplier before stocking a particular body of water, as some lakes (with poor water quality) have proved to be completely unsuitable for holding these animals. It is not uncommon for crayfish to literally climb out and walk away from an environment in which they are unhappy.

CONCLUSIONS

It is unlikely that crayfish will ever produce complete weed control over a range of waters, but occasionally, where lakes have become quite densely populated by these animals some level of control has been achieved. For this reason, and coupled with the fact that they might represent a small 'cash crop', they have been included in this chapter.

6 Chemical weed control

Weeds, whether in a garden or a farmed crop are often controlled using chemicals (herbicides). For terrestrial weeds there are a large number of herbicides available, with a complete range of activity. For example:

- HIGHLY SELECTIVE herbicides control a small number of specific weeds from particular crops;
- SELECTIVE herbicides are used for 'groups' of weeds, such as broadleaf weedkillers for lawns;
- NON-SELECTIVE (or broad-spectrum) herbicides which can indiscriminately kill a complete range of plants on contact.

Similarly, there is a range of herbicides (from selective to broad-spectrum) available for use on aquatic weeds. Unlike terrestrial weedkillers however, the number of chemicals which can be legally used on or near water is very limited and their use strictly controlled.

Because of the dangers of using pesticides in water supplies (which may be used for drinking water or irrigation), many countries have their own legislation regarding aquatic herbicide use. In the UK a small number have been cleared for use in or near water under what is known as the Food and Environment Protection Act (1984) Control of Pesticides Regulations (COPR).

The aim of these regulations is to advise on the correct use of approved herbicides with particular reference to farm irrigation, fisheries and public supplies. Although some of the information contained in these regulations is reproduced in this chapter, it is recommended that the document should be consulted in its entirety (see *Appendix 9*).

Fig 25 Selective chemical treatment with a knapsack sprayer

Role of the Regional Water Authorities

Water Authorities in the UK have the power to protect water quality under legislation; *eg.* the Prevention of Pollution Acts (1951, 1961), which control the input of any polluting matter (including herbicides) to inland waters. They are also responsible for the protection of fish under the Salmon and Freshwater Fisheries Act (1975). Therefore to ensure the safety of water courses for the public and for fish *consent must be obtained from the Regional Water Authority before aquatic herbicides can be applied.*

Deoxygenation

With the exception of some growth suppressors (which are not considered in this book) most chemical herbicides have been developed to kill plants. In the aquatic environment, the sudden death of large quantities of plant material, (particularly submerged varieties), frequently leads to severe deoxygenation with resultant fish kills (*Fig 26*).

With some herbicides (*eg* terbutryne), deoxygenation problems are compounded by the fact that shortly after treatment, the photosynthetic

Fig 26 Large quantities of weed left to die in an enclosed body of water can have disastrous results

pathways of the plants are blocked and oxygen production is stopped, yet the plants continue to utilise oxygen while respiring. Further (and more acute) deoxygenation continues when the plants die and are decomposed by micro-organisms.

The risk of deoxygenation is most acute in the summer when plant biomass and rate of respiration is at its highest, and the oxygen carrying capacity of the water is at its lowest due to the warm temperatures.

Most chemical treatments of submerged weeds are therefore performed during the spring when weed growth has just commenced and water temperatures are lower. Alternatively, if weed has to be treated later in the season, then only small selected areas should be treated at a time. Fish have the opportunity to move to untreated areas where dissolved oxygen levels are higher.

Choice of herbicide

Herbicides are generally expensive and potentially damaging to the environment, and therefore the most appropriate herbicide must be selected for each situation and used according to the manufacturer's instructions. This should ensure that the target weed is controlled as economically as possible and that the environment is safeguarded. Using inappropriate herbicides or using herbicides incorrectly is not only wasteful but can be extremely damaging.

Factors affecting herbicide choice will be:

- the weed species to be controlled. Target weeds must first be identified before selection can proceed;
- the timing of treatment. Herbicides which act through water are normally applied in the spring. Foliar sprays are usually applied in late summer;
- the use of water below point of application. With certain herbicides special care has to be taken to ensure that the water body is not being used for irrigation, livestock drinking, abstraction for domestic supply, fisheries, cress beds, *etc*. Special consideration should also be taken where susceptible crops are growing close to the water, which may be contaminated by spray drift;
- the velocity of the water. Some herbicides can be effective in flowing water whereas others can only be used in static situations.

There are only nine chemicals which have been given clearance for use in the UK on or near water. Some of these chemicals, however, have now been superseded by herbicides which offer a much wider range of weed control. Hence from the list of nine chemicals, only four are now commonly used for aquatic weed control. They are listed below:

Chemical name	*Trade name*	*Manufacturer*
Glyphosate	Roundup	**Monsanto**
	Spasor	**May & Baker**
Dichlobenil	Casoron (G/G-SR)	**ICI Agrochemicals Ltd**
Diquat	Midstream	**ICI Agrochemicals Ltd**
	Reglone	**ICI Agrochemicals Ltd**
Terbutryne	Clarosan	**Ciba-Geigy**

(A complete list of approved herbicides can be found in *Appendix 3*).

The susceptibility of common aquatic weeds to these herbicides can be determined from the following chart. Target weed identification is necessary before the chart can be utilised. Common weeds can be identified from *Section B* using either the quick guide (pages *68–70*) or the key (pages *106–109*).

Susceptibility of common aquatic weeds to approved herbicides

	Glyphosate	Dichlobenil	Diquat	Terbutryne
EMERGENT PLANTS				
Carex spp (sedges)	S	–	–	–
Glyceria maxima (reed sweet-grass)	S	–	–	–
Phalaris arundinacea (reed canary-grass)	S	–	–	–
Phragmites australis (common reed)	S	–	–	–
Schoenoplectus lacustris (club-rush)	–	–	–	–
Sparganium erectum (branched bur-reed)	–	–	–	–
Typha latifolia (bulrush)	S	–	–	–
Hippuris vulgaris (mare's-tail)	–	S	–	–
Rorripa nasturtium-aquaticum (water-cress)	S	MS	–	–
Sagittaria sagittifolia (arrowhead)	–	S	–	–
FLOATING PLANTS				
Lemna minor (duckweed)	S	–	S	S
Nuphar lutea (yellow water-lily)	S	–	–	–
Nymphaea alba (white water-lily)	S	–	–	–
Potamogeton natans (broad-leaved pondweed)	R	MS	MS	–
Polygonum amphibium (amphibious bistort)	S	–	–	–

SUBMERGED PLANTS

Callitriche stagnalis (water-starwort)	–	S	S	S
Ceratophyllum demersum (hornwort)	–	S	S	S
Elodea canadensis (Canadian pondweed)	–	S	S	S
Myriophyllum spp (water-milfoil)	–	S	S	S
Potamogeton crispus (curled pondweed)	–	S	S	S
Potamogeton pectinatus (fennel pondweed)	–	S	MS	S
Ranunculus spp (water-crowfoot)	–	S	S	S

ALGAE

Cladophora spp (cott)	–	–	MS	S
Enteromorpha intestinalis (bladder weed)	–	–	MS	S
Rhizoclonium spp (cott)	–	–	–	S
Spirogyra spp (cott)	–	–	MS	S
Vaucheria spp (cott)	–	–	–	MS

KEY

S = susceptible
MS = moderately susceptible
R/– = resistant/no information

Technical data for recommended herbicides

In the following section the technical data for each of the recommended herbicides is included to demonstrate the variation in application technique, dosage, timing, *etc*. The information included is by no means exhaustive; product labels and manufacturers technical literature must be consulted before any herbicide application is considered.

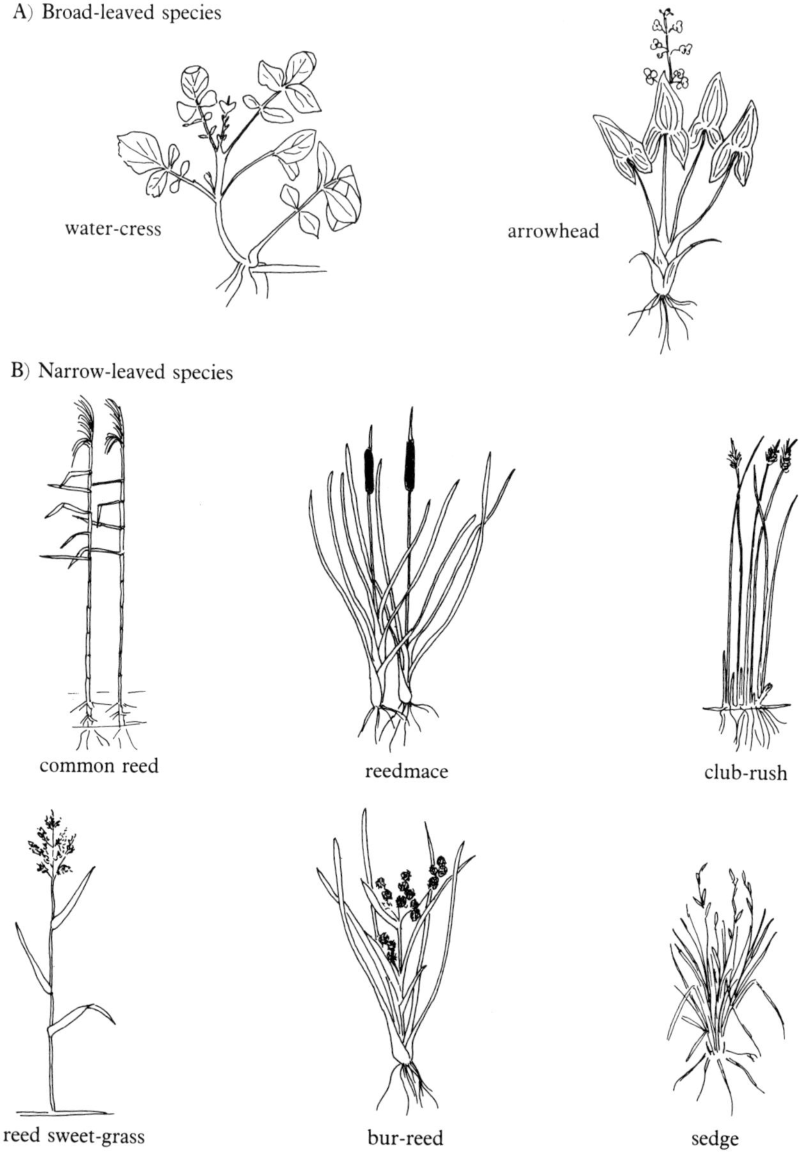

A) Broad-leaved species

water-cress   arrowhead

B) Narrow-leaved species

common reed   reedmace   club-rush

reed sweet-grass   bur-reed   sedge

Fig 27 Quick guide for identifying common emergent plants

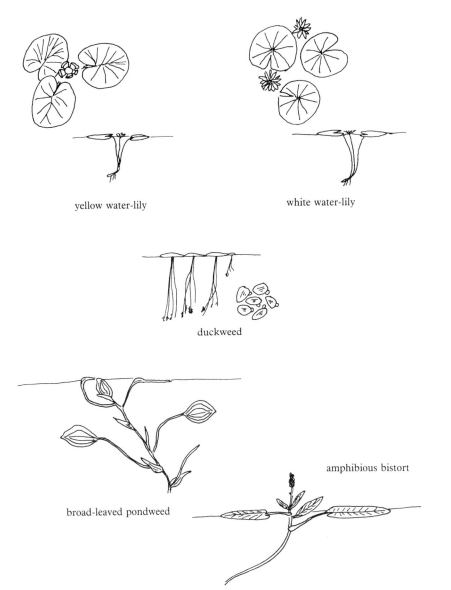

Fig 28 Quick guide for identifying common floating-leaved plants

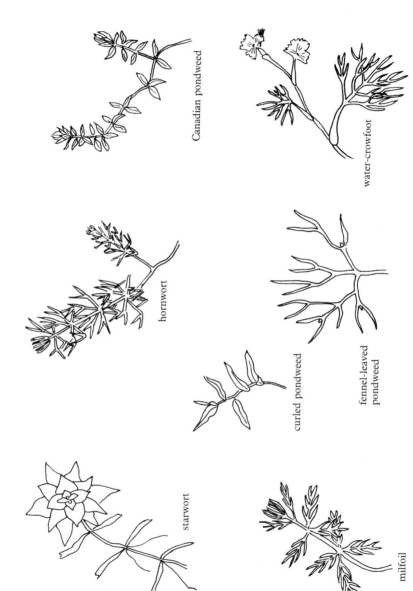

Fig 29 Quick guide for identifying common submerged plants

GLYPHOSATE

Products

Roundup (Monsanto) – active ingredient 359/L
Spasor (May & Baker) – active ingredient 360/L

General information

Glyphosate is an organophosphorus compound which is generally regarded as one of the most successful herbicides to be developed in recent years. It is used to control many of the more important emergent weeds (*ie* reeds, rushes and sedges) as well as floating-leaved plants such as lilies.

Formulation

It comes in liquid form which allows for easy dilution for application *via* knapsack, tractor or boat sprayer. Target weeds are sprayed with the recommended volumes. At least six hours dry weather is needed to ensure adequate foliar uptake.

Action

Glyphosate is a foliar acting herbicide, which means it is sprayed directly onto the target vegetation. The chemical is taken up by the plant leaves and translocated down to the root system where it acts. The effectiveness of the herbicide is dependent on how efficiently this translocation occurs. The timing of application is therefore extremely important and is determined by the development (size/age *etc*) of the target plants.

Timing and dosage

This herbicide is most effective when weeds are actively growing, particularly late in the season (mid-July – mid-September). Only one treatment per season will be necessary.

 'Grass' weeds (such as reeds, sedges *etc*) are usually treated in mid-August – mid-September (before die back) at 5 litres/hectare. Lilies need a higher concentration of 6 litres/ha and are best treated from

Fig 30 Treating emergents with glyphosate 1. tractor spraying

Fig 31 Treating emergents with glyphosate 2. boat spraying

mid-July – mid-August. By law the maximum permitted concentration allowed in water is 0.2 ppm of the active ingredient.

Symptoms

Affected weeds will normally exhibit discoloured/drying foliage within two–three weeks (lilies may take up to six weeks). In ideal growing conditions the action will be noticeably faster.

Regrowth should not occur the following year if the herbicide is applied correctly. Recolonisation, however, may begin from untreated margins *etc* which may warrant a new treatment after a few years.

Species susceptibility

Susceptible: *Carex* spp (sedges)
Glyceria maxima (reed sweet-grass)
Juncus effusus (soft rush)
Phalaris arundinacea (reed canary-grass)
Typha latifolia (bulrush)
Lemna minor (common duckweed)
Nuphar lutea (yellow water-lily)
Nymphaea alba (white water-lily)
Pteridium aquilinum (bracken)
Rumex spp (docks)
Phragmites australis (common reed)
Rorippa nasturtium-aquaticum (water-cress)
Catabrosa aquatica (whorl-grass)
Agrostis stolonifera (creeping bent)

Resistant: All submerged weeds and algae.

Important points (from manufacturers instructions)

Weeds should be fully developed (not cut) at application.
Avoid applying – to weeds starting to die back.
             – in windy conditions.
             – in high water volumes.
Direct spray to target weeds.
Avoid spray contact with crops or desirable plants.

Do not mix with other chemicals.
After application wash out sprayer.

Glyphosate, like all herbicides should not be swallowed or allowed to come in contact with skin or eyes. Protective clothing should be worn when using a knapsack sprayer.

Environmental effects

Under the COPR the maximum concentration of glyphosate permitted in water is 0.2 ppm. Good spraying technique should ensure that very little herbicide actually falls on the water. In terms of the environment, glyphosate is in fact a very safe herbicide to use as it is strongly absorbed by soil particles immediately on contact, and within 24 hours the residue levels in both soil or water are barely detectable. It has a very low mammalian toxicity and shows an extremely low bio-accumulation factor in fish. No significant adverse effects have been shown in mammals, birds, beneficial insects, and invertebrate fauna (Seddon 1981).

Deoxygenation problems are less likely when using glyphosate due to the late season application, the slowness of action and consequent death of the plants, and its more common use in marginal areas.

Fig 32 The result of an experimental treatment of lilies with glyphosate

DICHLOBENIL

Products
Casoron G   (6.75% ai)   (ICI)
Casoron G-SR   (20% ai)   (ICI)

General information

Dichlobenil is classed under the 'nitrile' group of herbicides. Both formulations have been primarily designed to control submerged weeds in still or slow-moving water (providing the flow rate does not exceed 90 metres per hour/25 cm per second).

Formulation

Casoron is produced in two granular formulations and contains 6.75% dichlobenil and Casoron G-SR contains 20%. Both types of granule control an identical spectrum of weeds.

The use of Casoron G-SR results in a better and more prolonged level of weed control than Casoron G, particularly with the 'moderately' susceptible species. The higher concentration also aids in labour saving for large water bodies. Where only low rates of application are envisaged, *ie* for small water bodies, then Casoron G is probably more economic and practical.

Use and mode of action

Casoron granules are applied directly into the water body where they sink into the mud and disintegrate. The active ingredient is then released and absorbed into the mud where it is uptaken by plant roots and transported round the plant. Susceptible weeds are ultimately killed.

This mode of action enables a partial treatment of submerged weeds to be undertaken, and enables localised weed control in discrete areas (*eg* for angling, boating, swimming *etc*). Casoron has been shown to be most effective against weeds with extensive root systems *eg* mare's-tail, milfoil, crowfoot and *Potamogeton* spp. Consequently, it is less effective against detached or free-floating weeds such as hornwort. Casoron is also not effective against the emergent reeds and rushes.

Timing and dosage

As with many other herbicides, the manufacturers recommend early applications, which treat weed as it begins to actively grow in March – May. This timing considerably reduces the possibility of killing large stands of weed and hence reduces the risk of deoxygenation. Once dense weed growth has become established, later applications can be considered as long as no more than 20% of the waterbody is treated.

Although the action of the herbicide is through the mud and hence root system of weeds, large water volumes can effectively dilute the active ingredient and hence water depth will determine treatment concentration. The following chart is taken from the manufacturer's instructions:

Water depth feet	Casoron G kg/ha	(lb/acre)	Casoron G-SR kg/ha	(lb 50/acre)
1	45	(40)	–	–
2	84	(75)	28	(25)
3	123	(110)	41	(37)
3+	150	(135)	50	(45)

For partial treatment the desired area will have to be determined so that the application can be calculated. For example, a 20 × 20 metre 'swim' for angling constitutes $400\,m^2$ or 0.04 ha. Depending on depth, this area would require the following weight of Casoron.

Water depth feet	Casoron G kg	(lb)	Casoron G-SR kg	(lb)
1	1.8	(4.0)	–	–
2	3.4	(7.5)	1.0	(2.2)
3	5.0	(11.0)	1.6	(3.5)
3+	6.0	(13.2)	2.0	(4.4)

Symptoms

Dichlobenil acts on the actively growing parts of plants which causes them to swell, turn brown and eventually die. It can therefore be particularly useful against young growing plants and hence early application is most effective.

Species susceptibility – the following list is quoted by the manufacturers

Susceptible:	*Callitriche stagnalis* (common water-starwort)
	Ceratophyllum demersum (rigid hornwort)
	Chara spp (stonewort)
	Elodea canadensis (Canadian waterweed)
	Equisetum fluviatile (water-horsetail)
	Equisetum palustre (marsh horsetail)
	Fontinalis antipyretica (willow moss)
	Glyceria fluitans (floating sweet-grass)
	Hippuris vulgaris (mare's-tail)
	Hydrocharis morsus ranae (frogbit)
	Hottonia palustris (water-violet)
	Lemna trisulca (ivy-leaved duckweed)
	Myriophyllum spp (water-milfoils)
	Potamogeton crispus (curled pondweed)
	Potamogeton pectinatus (fennel pondweed)
	Ranunculus spp (water-crowfoots)
	Rumex hydrolapathum (water dock)
	Sagittaria sagittifolia (arrowhead)
	Stratiotes aloides (water-soldier)
	Zanichellia palustris (horned pondweed)
Moderately susceptible:	*Alisma plantago-aquatica* (water-plantain)
	Berula erecta (lesser water-parsnip)
	Oenanthe spp (water-dropworts)
	Potamogeton natans (broad-leaved pondweed)
	Rorippa nasturtium-aquaticum (water-cress)
Moderately resistant:	*Nuphar lutea* (yellow water-lily)
	Nymphaea alba (white water-lily)
	Polygonum amphibium (amphibious bistort)
	Potamogeton lucens (shining pondweed)
	Sparganium erectum (branched bur-reed)
Resistant:	*Butomus umbellatus* (flowering-rush)
	Caltha palustris (marsh-marigold)
	Carex spp (sedges)
	Iris pseudacorus (yellow iris)
	Juncus effusus (soft rush)

Lemna minor (common duckweed)
Phalaris arundinacea (reed canary-grass)
Phragmites australis (common reed)
Schoenoplectus lacustris (common club-rush)
Typha latifolia (bulrush)
Most reeds, rushes and sedges, all species of algae except *Chara* spp and some floating species.

Important points

Even distribution is essential for good weed control and a mechanical applicator is regarded as essential by the manufacturers for large bodies of water. These can consist of manual applicators carried on the shoulders of the operator up to sophisticated motorised applicators for boat mounting.

The persistence of the chemical is usually one–two weeks depending on water quality and temperatures. The manufacturers recommend a 28 day period after treatment before water is used for irrigation purposes.

Precautions

The following safety precautions are recommended by the manufacturers.

WASH HANDS before meals and after work.
DO NOT DUMP surplus herbicide in water or ditch bottoms.
STORE IN ORIGINAL CONTAINER – tightly closed in a safe place.
EMPTY CONTAINER COMPLETELY and dispose of safely.
DO NOT USE TREATED WATER for irrigation purposes within *four* weeks of treatment.

DIQUAT

Products

Midstream	(Midox)	10% w/v viscous gel
Reglone	(ICI)	20% w/v solution

General information

This is a member of the pyridine group of herbicides. Diquat (and a similar product, paraquat) was developed in the late 1950s but is still considered important as a fast-acting broad spectrum herbicide. The chemical is water-soluble and is designed to treat submerged weeds and some algae. It is rapidly deactivated by soil/muddy water.

Formulation

Diquat is available for aquatic weed control in two contrasting forms. In its conventional liquid form (*ie* Reglone) it is diluted and used to treat entire sections or water bodies by conventional spray application. Recently a specially formulated viscous solution has been produced (Midstream) which, when applied to water via a special applicator, forms gelatinous 'strings'. These sink and stick to the foliage of submerged plants releasing the active ingredient, thereby acting as a contact herbicide.

Action

Reglone can be used in still water or slow-flowing water up to 90 m/hour. Midstream's formulation enables it to be used in quite fast-flowing water as the 'strings' become broken and lodge deep within the plant biomass.

This continued physical contact enables complete release of the active ingredient into the weed mass requiring control. In still water, the effect of Midstream is very localised and only those weeds in the immediate vicinity of the treated area will be affected so making it possible to treat particular areas (*eg* for angling swims). This 'partial treatment' is even effective against poorly rooted species such as hornwort and Canadian pondweed.

Timing/dosage

Reglone can be applied diluted or 'neat' at a concentration of 50 litres/ha/one metre depth = ($1\frac{1}{2}$ gal/acre/foot). For example, a one acre lake averaging 1 m (3 ft) depth will require $4\frac{1}{2}$ gallons, and a five acre lake averaging 1.5 m (5 ft) depth will require $37\frac{1}{2}$ gallons. Where weed infestation is heavy, treat only a quarter of the area at any one time. Successive sections should be treated at timed intervals (early applications when weed is just starting to grow actively in April/May will probably be safest).

Because of the very localised treatment which can be obtained with Midstream, timing is not critical, although the manufacturers suggest that most effective control can be obtained by treating in late spring or early summer when weeds are young and growth is active. Results may not be entirely satisfactory if treatment is made later in the season. This applies particularly to filamentous algae, where trials have shown that control is more effective if treatment is applied before the algal filaments reach the surface. Where infestations are heavy, specific areas should be treated at 14–21 day intervals.

Rate of use

Water less than 1 ft depth − 0.5 litre/100 m^2 = 3/4 pint/100 yds^2.
(At this rate one 5 litre container will treat 1,000 m^2).
Water greater than 1 ft depth − 1 litre/100 m^2 = 1$\frac{1}{2}$ pints/100 yds^2.
(At this rate one 5 litre container will treat 500 m^2).

The degree and persistence of control will depend on many factors, but one correctly timed application should give a full season's control for susceptible species.

Important note

Diquat (either formulation) is completely deactivated by mud. Treatment should be avoided in muddy water or delayed where sediment has been recently disturbed.

Application

Reglone can be applied either undiluted (by sub-surface injection) or by using conventional sprayers.

Midstream however *cannot* be applied with conventional spray equipment. A strengthened and modified sprayer has to be used to produce a high pressure 'jet' of the viscous liquid. This jet can be projected up to ten metres and as it hits the water a continuous gelatinous string is produced. Effective weed control will be produced a few metres greater than the area of surface treated due to sub-surface water movements. A 'downstream' effect in flowing water will also be apparent.

Details of the sprayer (and possibly hire of equipment) can be obtained from the supplier or from the sprayer manufacturer.

Calibration details can be found in manufacturers' instructions.

Species susceptibility

Susceptible: *Callitriche* spp (water-starworts)
                   Ceratophyllum demersum (rigid hornwort)
                   Elodea canadensis (Canadian waterweed)

| | *Lemna* spp (duckweeds) (Reglone only)
                          Myriophyllum spicatum (spiked water-milfoil)
                          Potamogeton crispus (curled pondweed)
                          Ranunculus spp (water-crowfoot)
                          Zannichellia palustris (horned pondweed)

Moderately susceptible: *Hottonia palustris* (water-violet) (Reglone only)
                          Potamogeton lucens (shining pondweed)
                          Potamogeton natans (broad-leaved pondweed)
                          Potamogeton pectinatus (fennel pondweed)
                          Sparganium emersum (unbranched bur-reed)
                          Cladophora spp ('cott' or blanket weed)
                          Enteromorpha intestinalis
                          Spirogyra spp

Resistant: *Hippuris vulgaris* (mare's-tail)
                          Nuphar lutea (yellow water-lily)
                          Nymphaea alba (white water-lily)
                          Polygonum amphibium (amphibious bistort)
                          Vaucheria spp ('cott' or blanket weed)
                          Most reeds and sedges

Important points

Maximum permitted concentration under the COPR = 2 ppm. Minimum interval between treatment and use of water for irrigation = 10 days (or until concentration in water drops below 0.02 ppm).

Safety

Diquat is a highly toxic chemical which is poisonous if swallowed and can be absorbed through the skin. Adequate protective clothing must therefore be worn during handling and application (*see manufacturer's instructions concerning safety procedure*).

TERBUTRYNE

Product

Clarosan (Ciba-Geigy) 1% w/w granule

General information

Terbutryne belongs to a group of chemicals known as the triazines. It is a widely used herbicide as it is active against a wide range of submerged weeds as well as many species of algae (blanket weed).

Formulation

The product name for terbutryne used for aquatic weed control is 'Clarosan'. It is produced as a granule and sold in 10 kg bags.

Action

The active ingredient is taken up by roots and leaves where it rapidly affects the process of photosynthesis. Plants stop growing almost immediately, but it may take two–four weeks before obvious signs of death appear.

Granules are spread evenly from the bank or a boat. In both cases a granule 'spreader' may facilitate even distribution. In situations with moving water, better results may be obtained by shutting off the flow for at least seven days.

If dense weed growth is to be controlled without causing deoxygenation, sections of the lake should be treated separately. Treating a quarter of the lake at a time is suggested, with an interval of at least 14 days between applications. This will allow fish to move into untreated areas if oxygen levels fall dangerously low. It is however important that the entire lake should be treated within a six to eight week period or weed controlling efficiency may be reduced.

Timing, dosage

For best results Clarosan should be applied when weed growth is active but before heavy infestations have built up. This is usually in April or May but may be as late as August.

Dose – the quantity of Clarosan used depends upon water volume.

Situation	Quantity
Susceptible plants and algae.	14 lbs/acre/foot ($5 \text{ kg}/100 \text{ m}^3$)

Moderately resistant plants and algae.   28 lbs/acre/foot
                                         (10 kg/100 m³)

Example 1 A two acre lake with an average depth of 4 ft will require 14 × 4 lb (56 lbs) of Clarosan to control susceptible plants.
Example 2 A one hectare lake with an average depth of 1.5 metres will require 75 kg of Clarosan.

Species susceptibility

Susceptible:	*Enteromorpha intestinalis* (bladder weed)
	Spirogyra spp ('cott' or blanket weed)
	Cladophora spp ('cott' or blanket weed)
	Rhizoclonium spp ('cott' or blanket weed)
	Ranunculus spp (water-crowfoot)
	Callitriche stagnalis (common water-starwort)
	Potamogeton crispus (curled pondweed)
	Potamogeton pectinatus (fennel pondweed)
	Elodea canadensis (Canadian waterweed)
	Myriophyllum spp (water-milfoils)
	Ceratophyllum demersum (rigid hornwort)
	Hottonia palustris (water-violet)
	Lemna spp (duckweed)
Moderately resistant:	*Vaucheria* spp ('cott' or blanket weed)
	Potamogeton natans (broad-leaved pondweed)
	Hippuris vulgaris (mare's-tail, only when fully submerged for at least three weeks after treatment).
	Nuphar spp (water-lilies)
	Nymphaea alba (white water-lily)
Resistant:	*Polygonum amphibium* (amphibious bistort)
	Emergent weeds such as reeds, sedges and rushes

Important points

There is evidence to suggest that terbutryne may be deactivated in very 'peaty' environments, *ie* when suspended solids are high. This was noted in some East Anglian drains.

Clarosan treated water may be used for irrigation purposes seven days after treatment.

Safety

Consult product label.
WASH hands before meals and after work.
DO NOT dump surplus herbicide in water or ditches.
KEEP in original container.
DISPOSE of empty container safely.
STORE in a cool dry place.
Maximum permitted concentration under the COPR = 0.1 ppm.

Conclusions

Herbicides make a considerable contribution to the armoury of fishery managers and drainage engineers for controlling aquatic weed. Their action extends to all types of weed control – from bankside vegetation to submerged weeds and algae, and in many situations herbicides may offer the only practical solution to a weed problem.

It must be said however that chemical weed control is not without its drawbacks. Some of the chemicals are quite hazardous to use (for the operator) and many could potentially do serious harm to the environment if treatment is not performed exactly to manufacturers specifications. Even a correctly applied treatment may create problems. It is often the case that an early treatment with say a 'submerged weed' killer will effectively control the problem only for another weed species (and often this may be blanket weed) to recolonise quite soon after the initial treatment. Growth of this second species is often quite rapid as it takes advantage of the large concentration of available nutrients which have been released into the water by the dead weed. This secondary weed problem may prove to be even worse than the original one, necessitating a further chemical treatment later during the season.

7 Special problems

Algae

The algae includes a very diverse spread of plants which range from microscopic single-celled forms as small as 1μm (1/1,000 mm) in diameter up to the multicellular seaweeds which can grow to 50 metres in length.

All forms of algae contain the green pigment 'chlorophyll', but in some species this may be masked by other pigments also contained within the cell. Thus algae can exhibit a very wide range of colours including red, brown, blue-green and even black.

In fresh water the largest forms are known as filamentous algae, commonly called blanket weed or cott. These species grow as long filaments which intertwine to produce dense mats.

Algae is often the primary food source for an aquatic environment and is an important producer of oxygen, even under the ice in winter. However, algae in all its various forms frequently causes weed problems, and these can be most difficult to rectify.

In fresh water there are three main groups of algae which will be considered separately in terms of their biology and control:
– single celled or unicellular algae;
– filamentous algae (blanket weed or cott);
– blue-green algae.

SINGLE CELLED ALGAE

Single celled or unicellular algae are microscropic in size and are found in a variety of shapes and forms. Examples of the more common groups found

Fig 33 Pond surface completely covered with filamentous algae

in fresh water include the 'motile' algae (such as *Chlamydomonas*) which possess flagella, enabling them to move freely throughout the water column. Non-motile varieties of algae are also common. They generally drift with water currents and adjust their buoyancy to find optimum light conditions. One particular group of non-motile algae are known as diatoms. They are recognised by a characteristic silica shell, which enables them to exist in a seemingly limitless range of shapes and forms.

Groups of unicellular algae are widespread throughout the aquatic environment and have managed to colonise almost any situation ranging from puddles and peatbogs to fast-flowing rivers.

They are of paramount importance to the food chain as they provide the primary food source for a whole host of small invertebrate animals which in turn provide food for fish. These algae are quick to colonise new environments which contain large concentrations of nutrients, ultimately colouring the water bright green and giving it a frequently undesirable 'pea soup' appearance. Such ponds will be highly productive in terms of the amount of planktonic animals they can support and hence the amount of food available for fish. Cyprinid fish farmers actually fertilise ponds to deliberately generate algal growth which indirectly provides unlimited food for their fish fry.

Algal blooms

In common with all plants, algae need light, nutrients and a suitable temperature to grow. When such conditions are optimal algal cells can undergo periods of very rapid multiplication which is termed a 'bloom'.

Such blooms occur at various times of the year but particularly in the spring. This is due to the fact that water temperature and light intensity are both increasing at this time of year. Under these conditions many species of algae are able to take advantage of the dissolved nutrients which have built up over the winter as a result of the decaying organic matter.

An interesting characteristic of algae is that most species tend to *store* more nutrients than they immediately require, and this frequently results in one or more of the essential nutrients suddenly running out. Algae will then rapidly die off over a period of a few days and sink down to the bottom mud. This phenomenon is termed a 'bloom collapse' and is characterised by the water changing from an opaque green/brown to virtually transparent over this short period of time. Frequently associated with a bloom collapse is a rapid fall in dissolved oxygen levels. This is due to the fact that there is little plant material actively producing oxygen, combined with an active utilisation of oxygen by the biological breakdown of the algal cells in the mud. This, of course, can be very dangerous in environments where there are high stocking densities of fish or where there is a presence of fish, such as trout, which are susceptible to low dissolved oxygen levels.

Because unicellular algae can multiply rapidly and utilise nutrients so efficiently, blooms can occur regularly in certain waters. Once the water has taken on the characteristic dense green colour, other plants can be shaded out leaving algae as the predominant species. Unicellular algae do not usually create problems for amenity use but blooms are not a desirable feature in situations which favour clear water such as a still water trout fishery or an ornamental lake.

Control

The control of a unicellular algae problem is rarely completely successful, as generally the symptoms are being treated rather than the underlying cause. This is particularly true when considering any form of chemical control, as algae populations, even when successfully treated, can very quickly recolonise.

Chemical control

Most species of algae are particularly sensitive to copper, and copper compounds are widely used in Europe and USA for controlling algal problems. Unfortunately copper sulphate has not been given clearance by the British Government to be used as a herbicide and its use is therefore prohibited in UK waters.

In the USA copper sulphate is frequently used. Muslin bags are filled with the crystals and towed behind boats in areas to be treated. It is, however, essential to achieve the desired concentration depending on the hardness of the water. Soft water requires up to 1 mg/litre whereas hard water (which deactivates the compound) requires much higher concentrations (5–12 mg/litre).

The great problem in using copper sulphate is that it is very toxic to fish and invertebrates (snails particularly). Domestic animals can also be harmed if they drink treated water. Serious environmental problems have frequently occurred by treatment based on wrongly calculated doses for particular water qualities. New and safer copper compounds have been developed for algal control (particularly a 'chelated' copper in the form of copper triethanolamine) but as yet these chemicals have not received final clearance for sale in the UK.

Existing copper compounds can be legally used in garden ponds which frequently suffer from algal problems. These compounds are quite successful and are sold as 'pond blocks', which can be readily bought at pet shops or garden centres.

Other forms of control are very limited. Undoubtedly the prevention or even a reduction in the amount of nutrients a lake receives or holds is the only long term answer, but sadly this is rarely a practical proposition. Other environmental measures such as shading with trees or the planting of lily beds will serve to reduce the light intensity reaching the water but this may only be effective in small lakes. There are some fish, such as the silver carp and some species of tilapia, which feed almost exclusively on algae, but it is very unlikely that they could effect any significant reduction in algal density, particularly in a UK climate.

FILAMENTOUS ALGAE (BLANKET WEED)

There are only a few species of filamentous algae which are commonly found in the UK, but when suitable environmental conditions prevail they can grow prolifically and create some of the most serious weed problems.

As their name suggests, this group of algae is composed of a 'chain' of identical cells which grows lengthways at the tips, or in some species by forming side branches. Reproduction is usually by fragmentation or under certain conditions by the formation of sexual or asexual spores.

Filamentous algae like unicellular species can take advantage of high concentrations of nutrients and grow rapidly either attached to the bottom or as free-floating forms. It is this latter form that causes most management problems by clogging channels and seriously affecting the amenity use of water (*ie* for angling, swimming, boating *etc*).

A factor which enhances the problems caused by blanket weed is that it produces large amounts of oxygen which become trapped within the filaments. This frequently causes quantities of it to rise and float on the surface.

Many still water trout fisheries which have high quality clear water ironically suffer from this problem the most. Sunlight can easily penetrate two or three metres to the lake bed and encourage the growth of the filamentous algae which is attached to stones or the bottom silt. As the rate of photosynthesis and hence oxygen production increases during the day sections break off and rise to the surface. Even very small quantities of blanket weed can cover a large surface area when it becomes 'inflated' with gas. This is particularly evident with species of *Spirogyra*, as it is covered with a slimy layer of musilage which greatly assists even the thinnest film of algae to remain floating. After a still, sunny day the result is often an entire lake surface covered in an algal scum making fishing almost impossible (*Fig 33*). A strong breeze may provide respite on some days by blowing the algae to the windward shore, but on a still day the algae will remain covering the surface. The only practical means of control is for the fishery manager to pull a seine net across the lake to create clear areas for fishing.

Another group of algae named *Cladophora* often floats to the surface in huge rafts where it continues to grow. Amenity use is seriously impaired and often large nutrient-rich lakes have huge quantities washed up on to the windward shore where it rots, smells and attracts numerous insects. *Cladophora* is probably the most common species found in the world and is regarded as the biggest nuisance.

In ecological terms blanket weed provides little benefit to the aquatic environment. Small (planktonic) animals cannot readily feed on this type of plant material (as they could singled celled algae) and hence there can be a reduced number of suitable invertebrate animals available within the food chain. This ultimately causes a reduced food supply for fish (particularly fry). Although some fish species, such as roach, will occasionally feed directly on blanket weed, as will crayfish, an excess within a fishery will ultimately lead to a reduced productivity in terms of fish growth.

There are many species of blanket weed which exists worldwide, but in general the species which commonly cause problems are restricted to the following:
Spirogyra, Ulothrix, Rhizoclonium, Stigeoclonium, Vaucheria, Hydrodicton and *Cladophora*.

Blanket weeds have high requirements for nutrients and light intensity, but how this group of algae can suddenly become dominant in a body of water over either unicellular algae or submerged plants is not clearly understood. The aquatic environment is far less stable and hence more complex than the terrestrial environment. Algae are generally short lived and are particularly sensitive to minute changes in the nutrient status and a host of other fluctuations in the system.

Often, blanket weed will flourish after a recent chemical or biological imbalance. This may occur in newly dug lakes (for the first two or three years), when a lake receives a nutrient rich input (*eg* runoff from surrounding land) or as a result of chemical weedkilling.

Blanket weed can rapidly take advantage of this situation and will frequently dominate other species of submerged plants. On occasions, however, it is the unicellular algae which is first to grow, and the resultant bloom can often shade out all other species of submerged plants, including the blanket weed.

Methods of control

Methods which can be used for controlling blanket weed are considered briefly in the following text. More detailed information regarding individual techniques can be found in *chapters 5* and *6*.

– *Chemical weed control*

Blanket weed can be controlled using copper sulphate (as described under unicellular algae) but, as previously commented upon, this chemical does not have clearance for use in water in the UK. There is, however, a permitted chemical named terbutryne (trade name

Clarosan) which can successfully kill most species of filamentous algae. The treatment works by producing the required concentration of herbicide in either the whole lake or if possible a section of lake. Unfortunately terbutryne also kills many other submerged weed species, and this is not always desirable.

The major drawback with chemical treatment is that when successful, the release of nutrients from the dead plant material quickly stimulates recolonisation, and within a couple of months the algae problem may return to pretreatment levels. Two or even three treatments a season may be necessary to maintain control, and this would prove quite expensive on a large body of water.

– *Biological control*
Grass carp (*Ctenopharyngodon idella*) will eat blanket weed but only after the more palatable species of submerged weed have been consumed. Common carp (*Cyprinus carpio*) can also control blanket weed especially if large strands of weed are first treated chemically. Lakes with muddy bottoms derive most benefit due to the shading effect of the 'coloured water' produced by the feeding fish. The end result, however, is a body of water coloured brown/green with unicellular algae which is frequently undesirable.

Crayfish will eat blanket weed but will not necessarily be selective against these species. Little information is available on the use of crayfish for weed control but they may have potential in a suitable environment.

– *Straw*
A novel and recent discovery is that conventional bales of straw can be used to effect some control over blanket weed. This phenomenon has been noticed when straw has been left in bales or spread thickly on areas of the lake bed, particularly in front of the inlet.

The reasons for this effect are as yet not clearly understood, but it is probable that the straw acts as a medium for large numbers of bacteria to colonise. The bacteria, which are quite harmless, compete with the algae for certain chemical nutrients, and eventually there is insufficient to enable the algae to develop.

BLUE-GREEN ALGAE

Blue-green algae are a most interesting group of organisms which although able to photosynthesise are quite different from other forms of algae or other plant species. The cellular structure of blue-green algae is much more primitive, and in evolutionary terms they represent a link between bacteria and green plants. To be precise they are not true algae.

They are widespread throughout the world and are found in a range of aquatic environments. Many common species are characterised by the fact they can produce a very obvious 'blue-green' colour in water, often described as looking like spilt paint. Occasionally they can accumulate near the surface where they form a scum and a very characteristic and extremely unpleasant smell.

Blue-green algae are found in both single celled or filamentous forms, and they frequently bloom in nutrient rich waters. Different species typically bloom at various times throughout the growing season from early spring to autumn. Like other forms of algae, water quality will determine their development, the critical factors being light, temperature, pH, nutrient concentration and the presence of soluble organic material. Some species grow well under low light intensity which gives them a competitive advantage, and could account for blooms later in the season.

Blooms which occur at the water surface form as a result in the change in buoyancy of the cells which are able to accumulate quantities of gas. On a still day, cells rise to the surface and produce the characteristic scum.

Importance of blue-green algae

Abundant growths of blue-green algae may be a source of considerable nuisance to man in terms of the management of reservoirs found in the more fertile lowland areas. Many large species physically clog the primary and fine sand filters used in the purification of drinking water.

Another important fact associated with the blue-green algae is that as part of their natural metabolism they release chemicals into the environment which have proved to be poisonous to fish, cattle and even man. These chemicals can also produce a most objectionable odour and taste to drinking water. Some grow attached to walls and pipes restricting flow, others can even corrode concrete and steel structures.

Toxins

The poisoning of farm animals (cattle, sheep *etc*) has been reported world wide, and attributed to the ingestion of lethal doses of toxic blue-green algae which have accumulated near the surface and along the shore of certain bodies of water. Three particular species (*Anabaena flos-aquae*, *Microcystis aeruginosa* and *Aphanizomenon flos-aquae*) appear to be responsible for all these reports.

When taken in a large enough quantity these toxins produce a neuromuscular blocking effect which leads to respiratory arrest.

It must be pointed out, however, that fatal poisoning is not common, as animals would have to ingest several litres of water containing a high concentration of the toxic blue-green algae to be seriously affected.

Control

Figures from the USA suggest that copper sulphate can be used to control blue-green algae blooms at concentrations of 0.5–1.0 ppm. This concentration is lower than those quoted for other algae as blue-green algae appear to be far more sensitive to copper than many other species. The use of copper sulphate is, however, restricted for use in UK waters as discussed earlier.

In large deep reservoirs (those which 'stratify' into different temperature layers during the summer), blue-green algae have been controlled by 'turning over' the water column by vigorous aeration. This technique breaks up the layers and brings up cooler nutrient rich water from greater depths causing a dramatic change in water quality. This frequently results in a change in the species of algae present. Usually species of green algae become dominant and the blue-green algae are replaced.

This form of control is regularly performed by Water Authorities in certain drinking water reservoirs under carefully controlled conditions and under the watchful eye of engineers and biologists. It could have detrimental effects upon the dissolved oxygen levels of the water (and hence the fish) and therefore may be an unsuitable technique for individual lake owners.

Apart from this rather specialised operation which serves to change rather than reduce algal populations, the limited number of techniques suggested for unicellular algal control should have equal application for the blue-green algae.

The total range of effective techniques for controlling any type of algae is, however, sadly very limited, and waters with algae problems are rapidly increasing in number.

Rivers

Rivers originate from many diverse sources, such as mountain runoff, artesian springs and lowland land drains. During their course to the sea, each river will be subjected to additional physical and chemical changes in response to the characteristics of the surrounding land. Each river type will therefore vary in terms of gradient, velocity, flow and nutrient status *etc*, and each river will change in character along its length. Consequently, the variety and quantity of plant life that each river can support will also be subjected to considerable variation.

Thus, as in static waters, the approach to aquatic weed control in rivers demands a variety of measures, depending on the characteristics and function of the stretch of water in question.

SLOW-FLOWING RIVERS

Slow-flowing lowland rivers, such as Fenland drains, disused canals *etc*, present some very serious problems regarding weed control. They become quite warm in the summer and frequently receive drainage water rich in nutrients; both factors encourage prolific growth of weed. The risk of flooding is ever present if this weed growth is left unchecked.

Such rivers which have very low velocities [below 0.2 m (7 ins)/sec] have similar plant communities as many still waters, thus many control measures may be common to both environments.

Weed control techniques (as for any body of water) will depend upon the nature and 'function' of the river. If, for example, a river is used for navigation and/or flood relief, then the priority should be to create a clear channel to allow both craft and flood water to pass unhindered. The most efficient action under these circumstances would be to clear the majority of aquatic vegetation with the exception of, say, a strip of marginals to prevent bank erosion. However, with an ever growing concern for conservation and to enhance angling potential, weed control needs to be far more discriminatory. In the past, engineers have usually decided that complete elimination of all weed has been the most efficient course of action. Happily, in response to pressure by angling and conservation bodies, a certain amount of compromise has been achieved and the habitat has benefited accordingly.

When considering slow moving rivers, the choice of weed control technique will usually be between mechanical devices (such as cutting boats, tractor mounted machinery *etc*) and chemicals.

Mechanical methods

Narrow channels are most easily managed using tractor mounted equipment, as long as the banks are negotiable (*Figs 34 and 35*). The reach of a

Fig 34 Clearing narrow streams with a hydraulically operated bucket

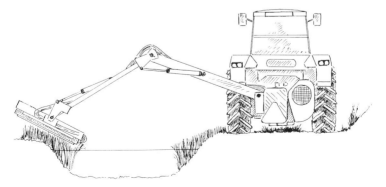

Fig 35 Emergent weed control using tractor-mounted flail mower (good bank access is essential)

hydraulic arm is dependent on the machine used but is usually in the range of 8–20 feet (*Fig 11*).

In severe cases of weeding up, conventional or 'Bradshaw'-type buckets can be used to lift out silt and submerged weed as well as the bankside emergents. In less serious cases tractor mounted cutting bars can be used to trim any type of weed to the required height. Cut weed should always be removed to prevent deoxygenation.

Chemical methods

Chemicals are regularly used for emergent weed control. Tractors can be used to spray long 'strips' of weed. If the banks lack ready access for machinery then hand spraying with back-packs can be used.

For submerged weed the recommended chemicals (see *chapter 6*) can be used but special consideration must be given to velocity of the water flow (follow manufacturer's application instructions).

Biological methods

Biological methods are not usually a prime consideration for rivers as the fish used may freely swim in and out of the stretch requiring control. Some drainage boards and Water Authorities are, however, experimenting with grass carp in stretches of rivers which are completely enclosed at either end by pump grills, mills *etc*, and with some very encouraging results.

Environmental methods

Shading from trees will greatly reduce plant growth, particularly in narrow rivers. Planting must, however, be concentrated on the south-facing banks (see *Fig 36*). *NB* If tractor mounted machinery is likely to be required on a stretch of water trees must be positioned in such a way as not to restrict access.

FAST-FLOWING RIVERS (with particular reference to chalk streams)

Unpolluted fast-flowing rivers may well have the potential to support populations of salmonids and they are frequently managed with the aim of producing a trout fishery.

It is well known that chalk streams provide the best environment for

Fig 36 The shading effects of trees planted on river banks

trout growth, but the very clear mineral rich water also provides a perfect environment for plant growth as well. The efficient management of chalk streams requires an altogether more professional approach to weed control than that of other rivers due to their considerable value as sport fisheries.

The large quantities of submerged weed found in chalk streams provides food and shelter for the enormous number of insects and other animals which inhabit the river. These in turn provide the basic diet of the trout.

The submerged plants are also necessary in providing 'cover' for the fish (so essential for trout) as well as contributing to the overall aesthetic appeal of the environment. If, however, the growth is allowed to become too prolific, fishing will quickly become impossible as the river will rapidly choke with weed. In extreme cases flooding will result.

Taking all these factors into consideration, it is clear that a certain balance needs to be achieved between providing food and shelter for fish, and weed clearance for angling and flood prevention.

As good quality trout fishing can generate a substantial income from a stretch of river, many riparian owners can afford to employ a full time riverkeeper. The keeper's primary function is to control plant growth in the river and on the bank, to produce optimum conditions for angling.

Traditionally most weed control in and alongside chalk streams is performed by mechanical methods, either by cutting with handtools (slashers, chain scythes *etc*) in the shallow sections and with weed cutting boats used on the long deeper sections (particularly on the Rivers Test and Itchen in Hampshire).

The use of these mechanical methods encourages a high degree of selectivity regarding weed control. Desirable plant species such as *Ranunculus* (water-crowfoot) are encouraged and trimmed, whereas less desirable species which root deeply and encourage silting (such as mare's-tail) may be cut right out.

Chemical weed control is rarely employed in chalk streams although it may have uses against the marginal species in overgrown side streams and feeders.

Weed cutting in shallow stretches of chalk stream over the years has led to the creation of some traditional 'patterns' which can greatly influence both the appearance and characteristic of the river. Two common cuts are described as follows:

- Checkerboard (*Fig 37*)
  Riverkeepers may argue as to the optimum ratio of plant cover to open water, but a 60:40 ratio in favour of plant cover is generally accepted as suitable in a good fishery. This permits fish to emerge from the shelter of a stand of plants into clear areas to feed.
- Cut and bar (*Fig 37*)
  The great attraction of this cut is that in addition to providing the accepted ratio between cover and clear areas, the plant 'bars' created

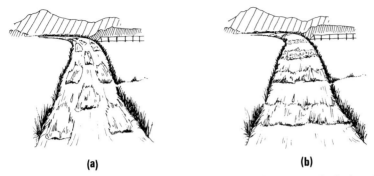

Fig 37 Common weed cutting patterns for a small chalk stream a. checkerboard b. cut and bar

have the effect of holding up the water level. Each 'bar' acts as a small weir which when created in a sequence has a significant effect on the overall water depth. This is particularly noticeable in small streams which have low summer flows and might otherwise not hold a head of fish.

Seasonal weed cutting

Over the years, riverkeepers have developed a rather seasonal approach to weed cutting and follow a basic strategy.

- Spring. As plants start to grow in the spring, patterns of cuts are established (*ie* checkerboard, bar *etc*). Undesirable species may be selectively cut right out and bank side emergents are shaped. All plant growth is encouraged on the outer bends to prevent erosion. Beds of cress may be encouraged to grow along the margins of unnecessarily wide stretches to 'narrow' the river and increase flow.
- Summer. When weed growth is at its peak the beds are trimmed to keep their shape, and to prevent the plants reaching the surface. The marginal emergents are kept short and all bank vegetation is kept trimmed along and behind paths to facilitate the angler's back cast (*Figs 38 and 39*).
- Autumn. Once the trout angling season has ended (October), shallow gravel areas are cleared of plant growth to provide suitable spawning areas. In other areas all plant growth is cut from the centre of the river

Fig 38 Emergent vegetation is trimmed and trees are carefully pruned to avoid interference with casting

Fig 39 The River Test at Leckford with well maintained bankside vegetation

leaving a strip of submerged and marginal weed along the edges. This cut is designed to encourage flood water to pass unimpeded along the centre of the river scouring away the deposited silt but leaving the banks protected (*Fig 40*).
- Winter. As plants are not growing in the winter months, riverkeepers usually turn their attention to bank control. Small feeder streams which have become overgrown may be dug out and overhanging branches pruned.

Weed cutting days

The time of year when weed growth is most prolific unfortunately coincides with the trout fishing season (April to October). The majority of weed cutting has therefore to be performed during this period. Unfortunately weed cutting activities create a number of problems for downstream anglers as the water becomes highly coloured and large quantities of cut weed continuously drift past (*Fig 41*). So, in an attempt to reduce the disturbance caused to anglers, specific weed cutting days (which are spaced throughout the summer) are agreed upon before the season commences, and angling is suspended during these periods.

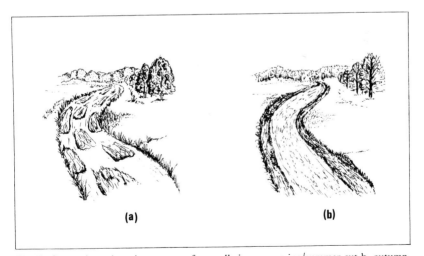

Fig 40 Seasonal weed cutting patterns for small rivers a. spring/summer cut b. autumn cut

Fig 41 Weed cutting in rivers can result in large quantities of weed affecting downstream activities

Section B
Identification, characteristics and control of common weeds

- Introduction *104*
- Keys for identifying problem plants *105*
   Emergent plants:
      Broad-leaved (*Key 1*) *106*
      Narrow-leaved (*Key 2*) *107*
   Floating-leaved plants (*Key 3*) *108*
   Submerged plants (*Keys 4 and 5*) *108, 109*
- Details of keyed species *111*

Introduction

This section will enable anyone with a little knowledge of plants to identify weeds which are highlighted as a problem in the text.

The key will guide the reader to a page reference for a detailed description of the plant which will confirm identification. (*Note*: if no page reference is given then there is no further description).

Each key section works in the form of a flow diagram and at each point a decision must be made and that route followed to the appropriate circled group of plants. In general, identification can be made entirely on leaf characteristics although in one case a group of plants have been classified on their characteristic flowers. In this special case there are details of leaf identification in the individual plant specification.

The key is made up of three main sections. Each section refers to the overall habitat of the plant.

Lastly, some plants may cross over from one main group to another at various growth stages or habitats. Where necessary these plants will occur on each key.

It should be noted that a plant or parts of a plant can take on a different form depending on the type of environment it is in. (For example, leaves may be longer in running water than still water). This feature will be highlighted in the individual plant specifications.

Keys for identifying problem plants

- Emergent: plants growing in margins or with major growth above water surface *pages 106, 107*

- Floating: plants with leaves floating on the water surface *page 108*

- Submerged: whole plants growing under the water surface *pages 108, 109*

BROAD LEAVED EMERGENTS

Simple leaves

Compound leaves

Leaves with a network of veins

Leaves with parallel veins

Rumex

Polygonum p 126

Sagittaria p 121

Alisma

Rorippa pp 119-20

Apium

Key 1 Broad-leaved emergents

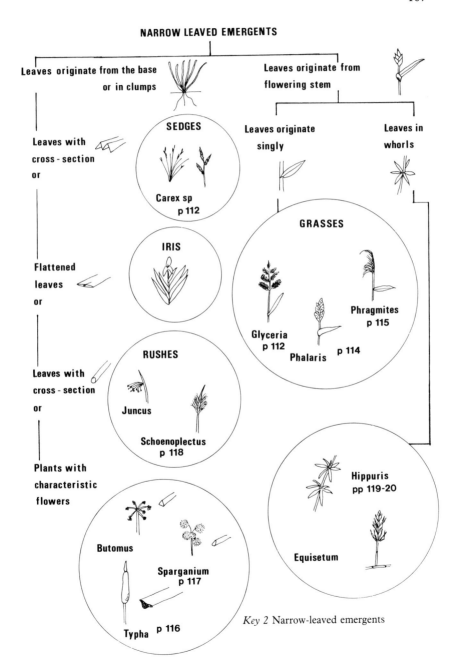

Key 2 Narrow-leaved emergents

FLOATING LEAVED PLANTS

SIMPLE LEAVES

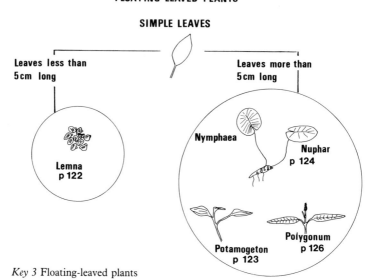

Leaves less than 5 cm long

Lemna p 122

Leaves more than 5 cm long

Nymphaea
Nuphar p 124
Potamogeton p 123
Polygonum p 126

Key 3 Floating-leaved plants

SUBMERGED PLANTS

Compound leaves

Simple leaves

Leaves arise individually

Leaves in whorls

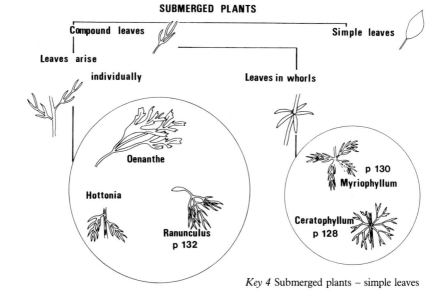

Oenanthe
Hottonia
Ranunculus p 132

Myriophyllum p 130
Ceratophyllum p 128

Key 4 Submerged plants – simple leaves

109

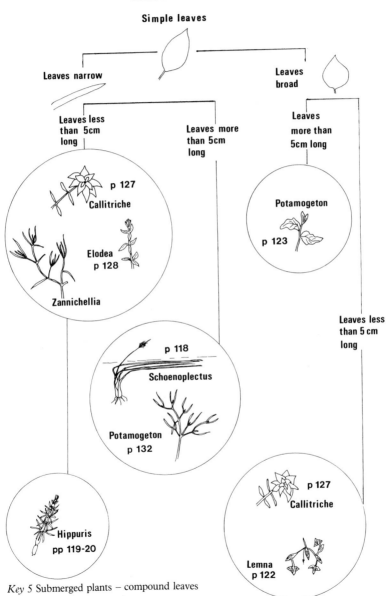

Key 5 Submerged plants – compound leaves

Details of keyed species

Emergent plants

SEDGES (*Carex* species)

Habitat
Sedges tend to be restricted to the very margins of rivers and lakes as they prefer marshy ground.

Botanical characteristics
There are many species within this group. They are all perennial with rhizomes. Plant heights range from 10–160 cm (4–5 ft). Leaves are characterised by having an 'M' shaped cross-section with triangular stems. Flowers are usually brown/black in colour.

Importance
Sedges can be responsible for siltation in lake margins *etc* but they are usually less vigorous than reedmace, reed *etc*. Some species only grow in the form of 'tussocks' which rarely cause problems of encroachment.

Control
Chemical – can be controlled with glyphosate.
Mechanical – efficient control can be produced with cutting but it has to be repeated twice a season.

REED SWEET-GRASS (*Glyceria* species)

Habitat
This true grass species is common in the margins of many lowland slow-flowing rivers and lakes.

Botanical characteristics
Glyceria maxima is the only truly emergent species. Plants grow up to 2 m ($6\frac{1}{2}$ ft) high, in up to 70 cm ($2\frac{1}{2}$ ft) depth of water forming dense luxuriant stands. Leaves are characterised by having boat-shaped tips. Flowers produced in July/August are large, branched and are generally a cream colour.

Importance
Glyceria is characterised by forming dense stands or occasionally large 'floating' rafts which steadily encroach. The dead vegetation breaks down much more readily than reedmace or reed.

Control
Chemical – can be controlled with glyphosate.
Mechanical – rafts of *Glyceria* can be frequently cut adrift to be winched or dragged ashore.

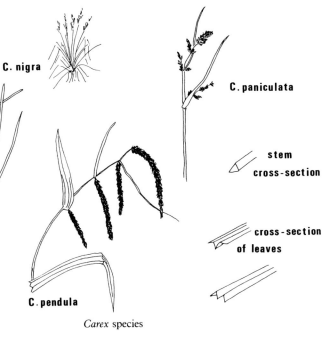

Carex species

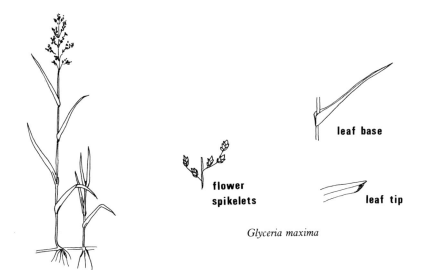

Glyceria maxima

Emergent plants

Phalaris arundinacea

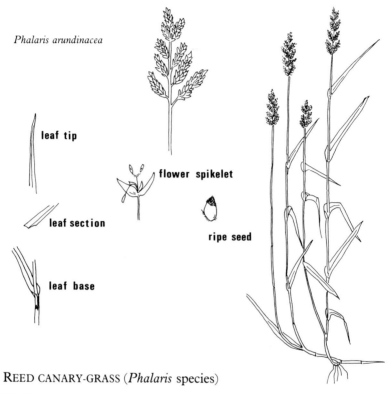

REED CANARY-GRASS (*Phalaris* species)

Habitat
Margins of rivers and lakes, occasionally producing pure stands.

Botanical characteristics
A perennial grass 0.5–2.0 m high (2–6 ft). It spreads by creeping underground rhizomes. It produces obvious flowers in June–August which are cream to reddish in colour. Leaves are blue-green with pointed tips.

Importance
Reed grass cannot tolerate shallow water and therefore does not encroach. It can however form dense stands along the margins.

Control
Chemical – readily controlled by glyphosate.
Mechanical – cutting will have to be repeated throughout a season but these species cannot tolerate grazing.

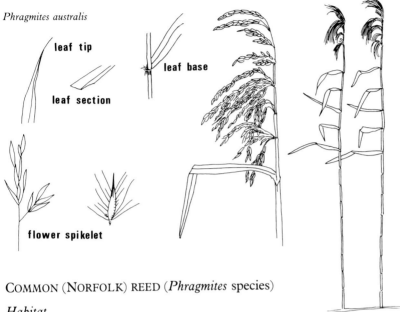

Phragmites australis

COMMON (NORFOLK) REED (*Phragmites* species)

Habitat
Reeds can grow prolifically in marshy environments as well as in open water up to 1.5 m (4 ft) deep. Also found in slow-flowing streams, but cannot tolerate fast flow.

Botanical characteristics
Reed is the tallest native grass in the UK which grows from 1–4 m (3–13 ft) high. It spreads by creeping rhizomes. Leaves are connected by a loose sheath so that the wind causes them all to point in the same direction. Flowers are produced in August and September and are a pale purple in colour.

Importance
The reed is prolific at 'encroaching' and like many of the emergents encourages silting due to the slow decomposition of the leaves and stems.

Control
Chemical – can be controlled by glyphosate.
Mechanical – cutting below the surface of the water can be quite effective but needs to be repeated. A single cut should be performed when the plant is growing vigorously during late July. Cattle may also keep the growth down.

Emergent plants

Typha latifolia

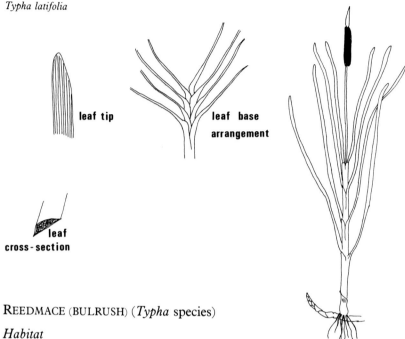

leaf tip

leaf base arrangement

leaf cross-section

REEDMACE (BULRUSH) (*Typha* species)

Habitat
Margins of lakes, rivers, ponds, marshes and bogs. Can form extensive stands in water up to a depth of 1.0 m (3 ft).

Botanical characteristics
Perennial, creeping rhizomes. Plants up to 2.5 m (8 ft) high. Leaves are blue-green in colour. Flowers between June and August and produces the characteristic red/brown spike.

Importance
Reedmace forms very dense stands and spreads quickly by sending up shoots from the rhizomes. The annual dead vegetation is slow to decompose which encourages rapid siltation and hence depth reduction below the stands. This greatly enhances its ability to encroach.

Control
Chemical – readily controlled by glyphosate.
Mechanical – cutting stems below surface encourages them to die back. Conventional cutting may have to be repeated several times a year. 'Pulling' is efficient in terms of complete control, but very slow.

Sparganium erectum

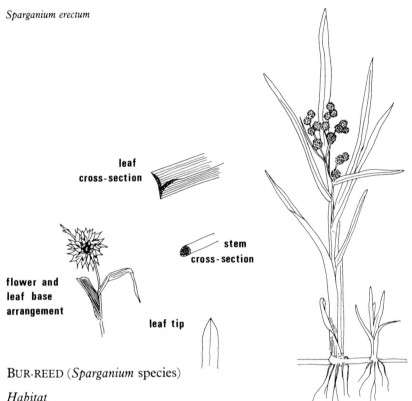

BUR-REED (*Sparganium* species)

Habitat
An emergent plant found in the margins of lakes and ponds frequently with reedmace and reed. It can form very dense stands in this zone. Other species can form floating leaves in some fast-flowing environments.

Botanical characteristics
This perennial spreads by a rhizome. Leaves are produced at the base of the stem and are triangular in section with an obvious keel running the complete length. The plant grows to some 1.5 m (5 ft) in height. Flowers are produced in June to August but the inflorescence is shorter than the leaves and are therefore not very obvious. When developed the ripe fruit is characteristic, and is described as having the appearance of a rolled-up hedgehog.

Importance
Bur-reed does form some dense stands but does not often encroach too far

into open water. The fruits are a nuisance to angling as they frequently entangle line. However, they form an important part of the diet of wildfowl during the winter.

Control
Chemical – bur-reed does not response readily to chemical control.
There is reference, however, to significant growth suppression being produced by the herbicide dalapon.
Mechanical – As with most emergents, cutting will control growth but bur-reed very quickly regrows, particularly early in the season, making frequent cutting necessary.

BULRUSH, CLUB-RUSH (*Schoenoplectus* species)

Habitat
An emergent plant which can produce dense stands in water depths of over 50 cms ($1\frac{1}{2}$ ft). It usually occurs on the inner fringe of the emergent zone as it can exist in deeper water than reed, reedmace *etc*. Club-rush is found in static water as well as moderate flowing streams. In faster water it forms floating and submerged leaves, sending up the stem and flower in an emergent form.

Botanical characteristics
This perennial plant has dark green cylindrical stems which grow up to 3 m (10 ft) tall and 1 cm thick at the base. The stems are leafless and are soft ('pithy') when squeezed. The rhizome is slow growing and spreads near the surface of the mud. Flowers are produced in June/July, red-/brown in colour and are clustered near the top of each stem.

Importance
This rush is very troublesome as it encourages siltation and binds the mud strongly around the roots making it difficult to dig out by hand. It is generally regarded as difficult to eradicate.

Control
Chemical – this plant does not lend itself too readily to chemical control. There is reference, however, to some growth suppression achieved with dalapon with a detergent wetting agent.
Mechanical – cutting will be effective during mid-season (July).

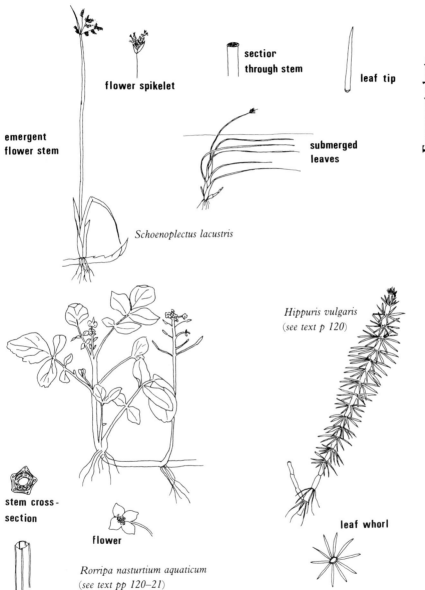

Emergent plants

Emergent plants

MARE'S-TAIL (*Hippuris* species) *see Fig p 119*

Habitat
This plant can exist as a totally submerged form in deep water, but in shallow situations it may emerge as much as 30 cm (1 ft) above the surface. It is found in a range of situations from static water to slow flowing rivers. It seems to prefer waters rich in calcium.

Botanical characteristics
A perennial plant with creeping rhizomes. The leaves are arranged in whorls and take on a soft (flaccid) characteristic below water, and a more rigid form above water. Flowers are small and inconspicuous.

Importance
Mare's-tail is rarely popular with fishery managers, as in both lakes and rivers it can form some quite dense stands. In clear lakes (particularly cool, calcareous environments) it can grow up from considerable depths *ie* 3–4 m (10–14 ft) giving the appearance of a dense forest of pine trees when viewed from above.

Control
Chemical – this plant will respond well to the granular formulation dichlobenil which it absorbs through the roots.
Mechanical – mare's-tail has a single stem which responds well to being cut and harvested. The plant does, however, regenerate quickly in the spring and early summer making further cuts necessary.

WATER-CRESS (*Rorripa* species) *see Fig p 119*

Habitat
Cress is found growing in the margins of rivers and streams, usually in pure (unpolluted) water. The plant can tolerate fairly swift flows.

Botanical characteristics
Cress spreads from a series of rhizomes. The stems grow erect, up to heights of 1 m (3 ft). The leaves are bright green (with a tinge of purple), smooth and shiny and have the characteristic 'hot' taste of cress. The flowers are small and white and are produced from May until September.

Importance
Cress is commonly found along the margins of chalk streams, but it is prolific and can easily encroach and cover small feeder streams. In larger rivers this characteristic is used to advantage by allowing the margins to

grow in and 'narrow' certain stretches of water which may have become too wide and slow.

Control
Chemical – like many true emergents, cress responds well to glyphosate treatment.
Mechanical – cress commonly forms rafts of vegetation in deeper water which can be readily raked ashore.

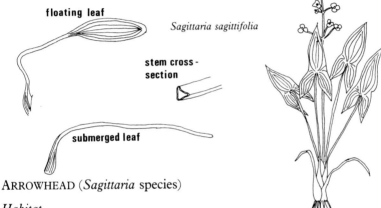

ARROWHEAD (*Sagittaria* species)

Habitat
Arrowhead is found growing in still/slow-flowing shallow water. It grows out of the water with very characteristic arrow-shaped leaves. (A ribbon-shaped submerged leaf may be formed in running water).

Botanical characteristics
The arrowhead overwinters by means of walnut sized tubers which are produced at the end of runners in the autumn. The tubers sprout in the spring producing ribbon-like submerged leaves, followed by long-stalked floating leaves and finally the typical emergent arrow-shaped leaves. They grow up to 90 cm (3 ft) high. A stem of white flowers is produced in July/August.

Importance
Arrowhead is not often a problem species but it can produce dense patches in shallow, silty water where it spreads rapidly.

Control
Chemical – arrowhead is readily controlled with the granular formulation dichlobenil which it absorbs through the roots.

Floating-leaved plants

DUCKWEED (*Lemna* species)

Habitat
Species of duckweed can be found on any still or slow-moving watercourse, particularly in large masses.

Botanical characteristics
Duckweed is composed of small leaf pads up to 0.5 cm (1/4 in) in diameter which bear one or several rootlets beneath. It reproduces almost exclusively by budding. Duckweed overwinters on the pond bottom in a 'dormant' state.

Importance
Duckweed multiplies at an alarming rate, creating complete cover over small ponds/canals *etc* in a matter of days, particularly if the environments are nutrient rich. When it completely covers water, duckweed creates problems by reducing oxygen exchange with the atmosphere and reducing light penetration to the water below. Both problems are extremely damaging to both animal and plant life in the aquatic environment.

Control
Chemical – although chemicals such as glyphosate, diquat and terbutryne will control duckweed, chemical treatments are difficult to apply effectively, particularly when the plant can regenerate so quickly.
Mechanical – shallow nets and booms can be used to reduce duckweed cover but regular treatment will be necessary.
Biological – young grass carp often have a preference for duckweed. If enough fish are feeding early in the season, duckweed can be prevented from becoming established.

overview of floating leaves

Lemna minor

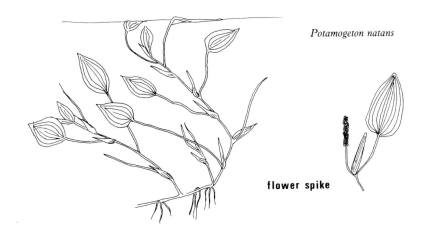

Potamogeton natans

flower spike

BROAD-LEAVED PONDWEED (*Potamogeton* species)

Habitat
This pondweed grows in slow-flowing and static water in water depths of up to 1 m (3 ft).

Botanical characteristics
This perennial plant is produced from a long, branched rhizome, the 'leathery' leaves are oval-oblong in shape up to 12 cm (5 in) long and are waxy on the upper surface. Submerged strap-like leaves may also be produced. Flower spikes are borne 8 cm (3 in) above the water in June to August.

Importance
Occasionally, this pondweed can produce a dense covering over the surface producing similar, but perhaps less acute, effects of shading on the environment as both species of lily. It is not often a popular weed on fisheries.

Control
Chemical – pondweed is relatively tolerant to hebicides but does exhibit a degree of susceptibility to dichlobenil and diquat. It is resistant to glyphosate.
Mechanical – cutting with chain scythes *etc* will provide effective control for a season once the plant has become established.

YELLOW WATER-LILY (*Nuphar* species)

Habitat
The yellow water-lily grows in slow-flowing and static water, in places which are sheltered from the wind. It grows up to depths of 2 m ($6\frac{1}{2}$ ft) generally in deeper water than the white lily.

Botanical characteristics
It is a perennial plant produced from long, stout rhizomes. Leaves are oval up to 40 cm (16 in) long and 30 cm (12 in) wide. The leaf stalk is triangular in section. Submerged leaves are thin and cabbage-like in appearance – under certain conditions these may be the only type produced. Flowers are produced in June to August. They are yellow and rise out of the water on stalks.

WHITE WATER-LILY (*Nymphaea* species)

Habitat
This lily grows in slow-flowing and static water, in water depths rarely greater than 1.5 m (5 ft).

Botanical characteristics
Very stout rhizomes produce round leaves up to 30 cm (12 in) in diameter. The leaf stalk is oval/round in cross-section. The production of submerged leaves is rare. Flowers are produced from June to August.

Importance (both species)
Beds of lilies give considerable aesthetic appeal to ponds and lakes. They offer cover and hold much invertebrate life which provides great attraction for fish. However, on occasion lilies can become so prolific that complete surface cover is produced. This is damaging for the environment below due to reduced oxygen exchange and complete shading as well as making angling and boating extremely difficult.
NB. Cultivars of the white water-lily form the range of coloured ornamental lilies popular for garden ponds *etc.*

Control (both species)
Chemical – all lilies can be treated with glyphosate to effect partial or total control.

Mechanical – due to the value of the white lily for ornamental purposes, rhizomes can be dug out of shallow water when leaves have died off. In deeper water chain scythes can produce reasonable control for a season (once the plant has become established).

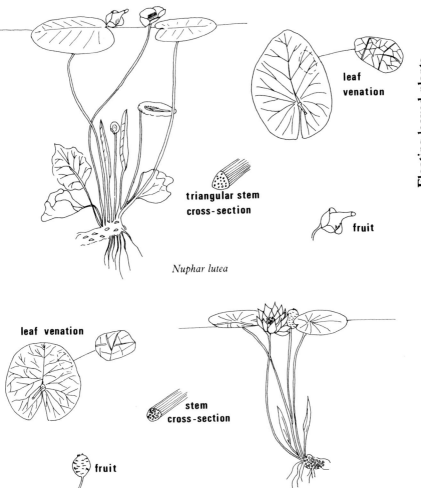

Nuphar lutea

Nymphaea alba

AMPHIBIOUS BISTORT (*Polygonum* species)

Habitat
An adaptable plant which is usually found in static or slow-flowing water.

Botanical characteristics
The plant has a creeping rhizome which produces stems up to 1 m (3 ft) long. The leaves are oval/oblong up to 10 cm (4 in) long by 4 cm (1½ in) wide. The characteristic flower spikes are pink and are borne above the water during July to September.

Importance
Bistort rarely grows in dense stands and thus does not produce the characteristic problems of lilies.

Control
Chemical – control can be achieved with glyphosate.
Mechanical – cutting with chain scythes *etc* will be effective for a season once the plant has become established.

Floating-leaved plants

Polygonum amphibium

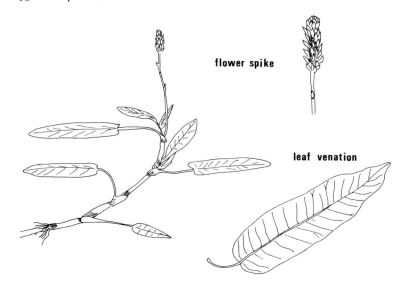

flower spike

leaf venation

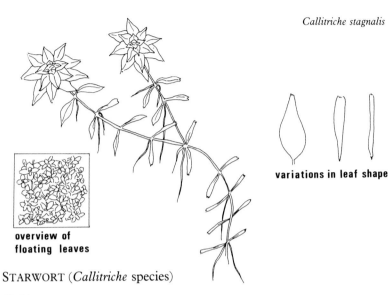

Callitriche stagnalis

variations in leaf shape

overview of floating leaves

STARWORT (*Callitriche* species)

Habitat
Species of starwort are found in a wide range of situations from static water and spring-fed ponds to moderate/fast-flowing rivers.

Botanical characteristics
The leaf shape of starwort varies according to species, but in general is up to 2.5 cm (1 in) long arranged in pairs on the stem. The colour of the plant is characteristically a *very* bright green. Often a terminal rosette of floating leaves is produced on the surface. Flowers are produced but are quite inconspicuous. Plants grow in clumps but seldom in very large stands. Individually, plants reach up to 1 m (3 ft) long.

Importance
In clear chalk streams the bright green cushions of starwort add considerable aesthetic appeal to the environment although the plant does hold back considerable amounts of silt. It is a good oxygenator and holds large numbers of invertebrates.

Control
Chemical – starwort is susceptible to dichlobenil, diquat and terbutryne.
Mechanical – beds of starwort can be easily cut with scythes *etc* but regrowth is rapid particularly in chalk streams.

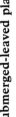

Submerged-leaved plants

CANADIAN PONDWEED (*Elodea* species)

Habitat
Canadian pondweed can be found in static to slow-flowing waters.

Botanical characteristics
This weakly rooted plant produces a much-branched stem which can grow up to 300 cm (10 ft) in length. It multiplies almost exclusively by fragmentation and winter buds. Old leaves and stems are dark green in colour whereas new growth is a lighter green.

Importance
Canadian pondweed, although a most prolific producer of dissolved oxygen, also produces some of the most dense stands of submerged aquatic weed. It causes considerable problems in waters all over the UK.

Control
Chemical – Canadian pondweed can be controlled with dichlobenil, diquat and terbutryne.
Mechanical – cutting will only serve to propagate the plant further, due to fragmentation.

HORNWORT (*Ceratophyllum* species)

Habitat
Found in static and slow-flowing water in the southern half of England. Rare in N. England, Wales and Scotland.

Botanical characteristics
Hornwort is a non-rooted plant that may float freely in the water or may anchor itself in the silt with portions of shoot. It can grow in quite dense masses reaching up to 90 cm (3 ft) in length. Propagation is by fragmentation of 'winter leaves' which can become detached from the shoot.

Importance
An important oxygenator which, in shallow water, can form dense beds which may interfere with angling *etc*.

Control
Chemical – hornwort can be readily controlled with dichlobenil, diquat and terbutryne.

 individual leaf

 whorled leaf arrangement

Elodea canadensis

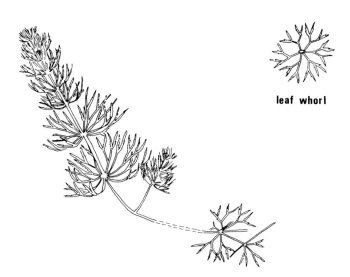

leaf whorl

Ceratophyllum demersum

Submerged-leaved plants

130

MILFOIL (*Myriophyllum* species)

Habitat
Milfoils are found in static and flowing waters. Some species have a preference for water rich in calcium.

Botanical characteristics
The submerged stems of the milfoils can grow up to 300 cm (10 ft) in length, bearing soft feather-like leaves. Like hornwort, this plant does not have a true root, but a modified 'creeping' stem by which it is anchored. Flowers are borne above water on a spike up to 20 cm (8 in) long. Some species produce specialised winter buds during September and October.

Importance
Milfoils can grow in fairly dense stands but the plant itself is quite soft and does not cause such problems as Canadian pondweed.

Control
Chemical – Milfoil can be readily controlled with dichlobenil, diquat and terbutryne.

Submerged-leaved plants

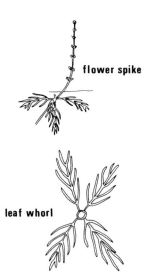

flower spike

leaf whorl

Myriophyllum spicatum

CURLED PONDWEED (*Potamogeton* species)

Habitat
Curled pondweed is found in still and flowing water.

Botanical characteristics
This plant has a thin, highly branched rhizome which forms dense mats in the silt layer. The leaves of the plant have an obvious 'crinkled' form which take on a reddish-green coloration. The plant produces winter buds and brownish coloured fruits close to the water surface. Some species can produce very long stems (up to 6 m/18 ft).

Importance
Curled pondweed, in certain situations, grows prolifically, frequently forming dense stands. There are a range of closely related species which can adapt to most environments.

Control
Chemical – curled pondweed is readily controlled with dichlobenil, diquat and terbutryne.

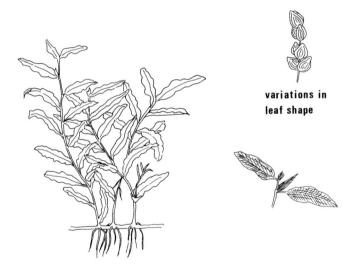

variations in leaf shape

Potamogeton crispus

FENNEL-LEAVED PONDWEED (*Potamogeton* species)

Habitat
This plant occurs in still, flowing and even brackish water.

Botanical characteristics
Fennel-leaved pondweed produces narrow thread-like leaves from a much-branched stem. The plant is rooted and can produce stems up to 2 m ($6\frac{1}{2}$ ft) long of a dull green colour. Winter buds and fruits are also produced.

Importance
Along with some closely related species, the plant can on occasion form dense stands particularly in shallow water.

Control
Chemical – fennel-leaved pondweed can be controlled with dichlobenil and terbutryne. It is less susceptible to diquat.

WATER-CROWFOOT (*Ranunculus* species)

Habitat
There are a large number of species of water-crowfoot which have adapted to live in a range of aquatic habitats, from small ponds to fast-flowing rivers.

Botanical characteristics
These rooted plants produce two characteristic leaf forms. The submerged leaf is divided into hair like filaments, whereas the floating leaves are lobed and rounded similar to the common buttercup. The flowers are produced in May to August once the plant has produced its floating leaves. The flowers are white with a yellow centre.

The plant overwinters from root stems in the mud. Under suitable conditions the plant can grow extremely fast, forming dense stands of intertwining stems over 3 m (10 ft) in length.

Importance
Crowfoot is without doubt the submerged plant most highly favoured by riverkeepers. It holds a large amount of insect food and will shelter fish but is not so dense around the lower stem to encourage much silting. It can, however, produce very dense stands of floating plant material if left unchecked.

Control
Chemical – crowfoot is readily controlled with dichlobenil, diquat and terbutryne, in still waters.
Mechanical – beds of crowfoot can be readily controlled with hand tools in rivers.

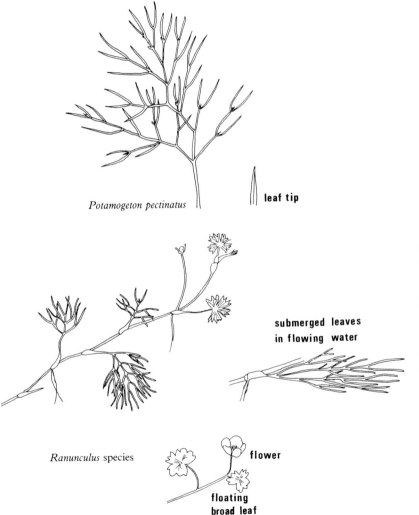

Potamogeton pectinatus

leaf tip

Ranunculus species

submerged leaves in flowing water

flower

floating broad leaf

Submerged-leaved plants

1 Cladophora
Branched filaments are up to 0.1 mm wide. The plant mass is often large and rough to the touch. It is found in running water (where it attaches to sticks and stones) or grows in tangled masses in lakes and ponds.

2 Rhizoclonium
Is very similar in both appearance and structure to *Cladophora* but the filaments are not branched.

3 Spirogyra
The filaments of *Spirogyra* are unbranched and are characterised by forming a surrounding slimy (mucilaginous) layer which is quite obvious to the touch. Under microscopic examination the plant is identified by the fact that the chloroplasts from a distinct, continuous spiral structure within the cell.

4 Vaucheria
A common, bright green filamentous algae of static and slow-flowing water. Individual filaments may be up to 0.2 mm wide. There is some branching but the plants are characterised by the fact that there are no cross walls.

5 Ulothrix
Unbranched filaments, each cell contains characteristically only one chloroplast which forms an incomplete band.

6 Stigeoclonium
Very similar to *Ulothrix* with band-shaped chloroplasts but with branched filaments. Softer to the touch than *Cladophora*.

7 Hydrodictyon
Net-like colonies common in slow flowing water. Well grown nets are large and conspicuous.

8 Enteromorpha
Has a unique hollow, tubular structure up to 1 cm wide and 15 cm long. It is so named because in water it has the appearance of an intestine! It occasionally forms conspicuous floating masses on the surface of slow rivers and canals. It may be abundant in brackish water.

Filamentous algae

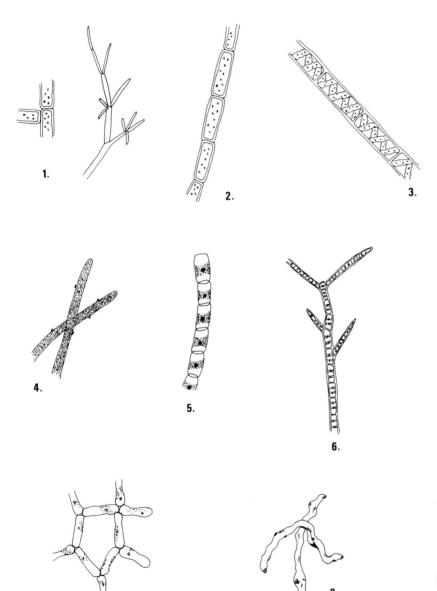

Filamentous algae

Section C

Appendices

1 Comparative costs *137*
2 Addresses of specialised equipment firms *138*
3 The complete list of approved chemicals *140*
4 Water Authorities in England and Wales *141*
5 River Purification Boards/Island Councils, Scotland *145*
6 Water pollution control in Northern Ireland *146*
7 Other helpful bodies *146*
8 Sprayers *147*
9 Further reading *149*

1 Comparative costs

It is not conventional to include current costs of equipment in text books as prices fluctuate considerably over quite short periods of time. However, to give the reader an idea of comparative costs between some of the techniques discussed in the text, the following table has been included.

The figures were produced by Dr Alan Frake of Wessex Water Authority in an effort to place the costs of large-scale weed control into perspective over the short term and a four year period.

	£/hectare	
	For first year	Over 4 years
Flowing water (drainage channels)		
Machine-mounted buckets	£2000	£8000
Weed cutting boats	£800	£3200
Herbicides	£500	£2000
Shading (black plastic)	£1000	£1000
Shading (trees)	£4000	£4000
Still water		
Herbicides	£525	£2100
Shading (black plastic)	£1000	£1000
Grass carp (at 250 kg/ha)	£1500	£1500

2 Addresses of specialised equipment firms in the UK

WEED CUTTING BOATS

John Wilder (Engineering Ltd)
Highercroft Works, Wallingford, Oxon. OX10 9AR
Tel: (0491) 37700

J. Bradshaw (Ltd)
New Lane, Stibbington, Peterborough, PE8 6LW
Tel: (0780) 782621

Liverpool Waterwitch Marine Engineering Co Ltd
74 Winifred Lane, Aughton, Ormskirk, Lancs L39 5DL
Tel: (0695) 422686

CHAIN SCYTHES

T. & J. Hutton Ltd
Dept T.S. Phoenix Works, Ridgeway, Eckington, Nr. Sheffield
Tel: (0246) 2088

HERBICIDE MANUFACTURERS

Ciba-Geigy Agrochemicals
Whittlesford, Cambridge CB2 4QT
Tel: (0223) 833621

Monsanto PLC
Agricultural Division, Thames Tower, Burleys Way,
Leicester LE1 3TP
Tel: (0533) 20864

May & Baker
Dagenham, Essex RM10 7XS
Tel: (01 592) 3060

ICI
Woolmead House, Bear Lane, Farnham, Surrey GU9 7UB
Tel: (0252) 724525

Fish Farms (Grass Carp/Mirror Carp)

Hampshire Carp Hatcheries
5 Segars Lane, Twyford, Hampshire
Tel: (0962) 712829

Fishers Pond Fishery
Fishers Pond, Colden Common, Hampshire
Tel: (0703) 694412

Humberside Fisheries Ltd
Cleaves Farm, Skerne,
Driffield, Humberside
Tel: (0377) 43613

Crayfish suppliers

Kingcombe Crayfish
St Francis Farm, Hooke, Beaminster, Dorset DT8 3NX
Tel: (0308) 862711

Michael Brown, Thorny, Langport, Somerset TA10 0DR
Tel: (0458) 215520

Two Lakes Fishery, Crampmoor, Romsey, Hampshire SO21 9BA
Tel: (0794) 512468

3 Complete list of products approved for use in or near water under the UK agricultural chemicals approval scheme

Chemical	Safety interval before irrigation	Approved products	For control of
asulam	nil	Asulox	Bracken and docks on banks beside water
chlorthiamid	4 weeks	Prefix	Some floating and submerged weeds
2,4-D amine	3 weeks	Chipman 2,4-D Dormone Fernimine	Emergent broad-leaved weeds and weeds on banks
dalapon	5 weeks	B.H. Dalapon Dow Dalapon SDS Dalapon	Reeds and similar emergent weeds
dichlobenil	2 weeks	Casoron G Casoron G-SR	Some floating submerged weeds
diquat	10 days	Reglone Midstream	Some floating and submerged weeds and algae Submerged weeds
fosamine ammonium	nil	Krenite	Deciduous trees and shrubs on banks beside water
glyphosate	nil	Roundup Spasor	Water-lilies, reeds and emergent weeds
maleic hydrazide	3 weeks	Regulox-K Vondalhyde-K	Suppression of grass growth on banks beside water
terbutryne	7 days	Clarosan	Some floating and submerged weeds and algae

(Take from 'Guidelines for the use of herbicides on weeds in or near watercourses and lakes'. MAFF Booklet B2078)

4 Water Authorities in England and Wales

ANGLIAN WATER AUTHORITY

Chief Scientist
Ambury Road
Huntingdon PE18 6NZ
Tel: Huntingdon (0480) 56181

Divisional Organisation

CAMBRIDGE
Divisional General Manager
Great Ouse House
Clarendon Road
Cambridge CB2 2BL
Tel: Cambridge (0223) 61561

COLCHESTER
Divisional General Manager
The Cowdray Centre
Cowdray
Colchester CO1 1BY
Tel: Colchester (0206) 69171

LINCOLN
Divisional General Manager
Waterside House
Waterside North
Lincoln LN2 5HA
Tel: Lincoln (0522) 25231

NORWICH
Divisional General Manager
Yare House
62/64 Thorpe Road
Norwich NR1 1SA
Tel: Norwich (0603) 61561

OUNDLE
Divisional General Manager
North Street
Oundle
Peterborough PE8 4AS
Tel: Oundle (0832) 73701

NORTHUMBRIAN WATER AUTHORITY

Assistant Director
Scientific Services
Northumbria House
Regent Centre
Gosforth
Newcastle-upon-Tyne
Tel: Gosforth (0632) 843151

Divisional Organisation

NORTHUMBERLAND AND TYNE
Divisional General Manager
Northumbria House
Town Centre
Cramlington NE23 6UP
Tel: Cramlington (0670) 713322

TEES
Divisional General Manager
Trenchard Avenue
Thornaby
Stockton-upon-Tees TS17 0EQ
Tel: Stockton-upon-Tees (0642) 62216

WEAR
Divisional General Manager
Wear House
Abbey Road
Pity Me
Durham DH1 5EZ
Tel: Durham (0385) 44222

NORTH WEST WATER AUTHORITY
Director of Planning
Dawson House
Great Sankey
Warrington WA5 3LW
Tel: Penketh (092572) 4321

Rivers Division
Manager
New Town House
Buttermarket Street
Warrington WA1 2QG
Tel: Warrington (0925) 53922

SEVERN TRENT WATER AUTHORITY
Director of Technical Services
Abelson House
2297 Coventry Road
Sheldon
Birmingham B26 3PU
Tel: Birmingham (021) 743 4223

Divisional Organisation

AVON
Divisional Manager
Avon House
12E De Montfort Way
Cannon Park
Coventry CV4 7E
Tel: Coventry (0203) 416510

DERBY
Divisional Manager
Raynesway
Derby DE2 7JA
Tel: Derby (0332) 61481

LOWER SEVERN
Divisional Manager
Southwick Park
Gloucester Road
Tewkesbury GL20 7DG
Tel: (0684) 294516

LOWER TRENT
Divisional Manager
Mapperley Hall
Lucknow Avenue
Nottingham NG3 5BN
Tel: Nottingham (0602) 608161

SOAR
Divisional Manager
Leicester Water Centre
Gorse Hill
Anstey
Leicester LE7 7GZU
Tel: Leicester (0533) 352011

TAME
Divisional Manager
Tame House
156/170 Newhall Street
Birmingham B3 1SE
Tel: Birmingham (021) 2331616

UPPER SEVERN
Divisional Manager
Shelton
Shrewsbury SY3 8BJ
Tel: Shrewsbury (0743) 63141

UPPER TRENT
Divisional Manager
Trinity Square
Horninglow Street
Burton-on-Trent DE14 1BL
Tel: Burton-on-Trent (0283) 44511

SOUTHERN WATER
AUTHORITY
Director of Administration
Guildbourne House
Chatsworth Road
Worthing BN11 1LD
Tel: Worthing (0903) 205252

Divisional Organisation

HAMPSHIRE
Divisional Manager
Otterbourne Waterworks
Otterbourne
Winchester SO21 2DP
Tel: Twyford (0962) 713622

ISLE OF WIGHT
Divisional Manager
St. Nicholas
58 St. John's Road
Newport PO30 1LT
Tel: Newport (0983) 52661

KENT
Divisional Manager
Luton House
Capstone Road
Chatham ME5 7QA
Tel: Medway (0634) 46655

SUSSEX
Divisional Manager
Falmer
Brighton BN1 9PY
Tel: Brighton (0273) 606766

SOUTH WEST WATER
AUTHORITY
Principal Scientific Officer
Scientific Services
Peninsular House
Rydon Lane
Exeter EX2 7HR
Tel: Exeter (0392) 219666

THAMES WATER AUTHORITY

Manager Environmental Services
Nugent House
Vastern Road
Reading RG1 8DB
Tel: Reading (0743) 593333

WELSH WATER AUTHORITY

Chief Executive
Cambrian Way
Brecon
Powys LD3 7HP
Tel: Brecon (0874) 3181

Divisional Organisation

NORTHERN
Divisional Manager
Divisional Headquarters
Penrhosgarnedd
Bangor LL57 2EQ
Tel: (0248) 351144

SOUTH EASTERN
Divisional Manager
Divisional Headquarters
Pentwyn Road
Nelson
Treharris CF46 6LY
Tel: Nelson (0443) 450577

SOUTH WESTERN
Divisional Manager
Divisional Headquarters
Hawthorn Rise
Haverfordwest SA61 1QP
Tel: Haverfordwest (0437) 4581

WESSEX WATER AUTHORITY

Chief Executive
Wessex House
Passage Street
Bristol BS2 0JQ
Tel: Bristol (0272) 290611

Divisional Organisation

AVON AND DORSET
Divisional Director
Divisional Headquarters
7 Nuffield Road
Poole BH17 7RL
Tel: Poole (02013) 71144

BRISTOL AVON
Divisional Director
Divisional Headquarters
P.O. Box 95
Quay House
The Ambury
Bath BA1 2YP
Tel: Bath (0225) 313500

SOMERSET
Divisional Director
Divisional Headquarters
P.O. Box 9
Bridgewater House
King Square
Bridgewater TA6 3EA
Tel: Bridgewater (0278) 57333

YORKSHIRE WATER AUTHORITY

Chief Scientific Officer
West Riding House
67 Albion Street
Leeds LS1 5AA
Tel: Leeds (0532) 448201

Divisional Organisation

CENTRAL
Divisional General Manager
Spenfield
182 Otley Road
West Park
Leeds LS16 5PR
Tel: Leeds (0532) 781313

NORTH AND EAST
Divisional General Manager
32/34 Monkgate
York YO3 7RH
Tel: York (0904) 642131

SOUTHERN
Divisional General Manager
Castle Market Building
Exchange Street
Sheffield S1 1GB
Tel: Sheffield (0742) 26421

WESTERN
Divisional General Manager
P.O. Box 201
Broadacre House
Vicar Lane
Bradford BD1 5PZ
Tel: Bradford (0274) 306063

5 River Purification Boards and Island Councils (with R.P.B. responsibilities) in Scotland

The Director and River Inspector
Clyde River Purification Board
Rivers House
Murray Road
East Kilbride
Glasgow G75 0LA
Tel: East Kilbride (03552) 38181/6

The Director
Forth River Purification Board
Olinton Dell House
West Mill Road
Colinton
Edinburgh EH13 0NX
Tel: Edinburgh (031) 441 4691

Director and River Inspector
Highland River Purification Board
Strathpeffer Road
Dingwall IV15 9QY
Tel: Dingwall (0349) 62021

The River Inspector
North East River Purification Board
Woodside House
Persley
Aberdeen AB2 2UQ
Tel: Aberdeen (0224) 696647

The Director
Solway River Purification Board
Rivers House
Irongray Road
Dumfries DG2 0JE
Tel: Dumfries (0387) 720502

The Director and River Inspector
Tay River Purification Board
3 South Street
Perth PH2 8NJ
Tel: Perth (0738) 27989

The Director and River Inspector
Tweed River Purification Board
Burnbrae
Mossilee Road
Galashiels TD1 1FA
Tel: Galashiels (0896) 2425

ISLAND COUNCILS

Director of Engineering and Technical Services
Orkney Island Council
Council Offices
Kirkwall KW15 1NY
Tel: Kirkwall (0856) 3535

Director of Design and Technical Services
Shetland Islands Council
92 St. Olaf Street
Lerwick ZE1 0ES
Tel: Lerwick (0595) 3535

Director of Engineering Services
Western Isles Islands Council
Council Offices
Sandwick Road
Stornoway PA87 2BW
Tel: Stornoway (0851) 3773

6 Water pollution control in Northern Ireland

Water Pollution Control Branch
Department of the Environment for Northern Ireland
Stormont
Belfast BT4 3SS
Tel: Belfast (0232) 63210

7 Other helpful bodies

Dr Barratt, Aquatic Weed Research Unit, Sonning Farm, Charvil Lane, Sonning-on-Thames, Reading RG4 OTH Tel: (0734) 690072

Freshwater Biological Association, River Lab, East Stoke, Wareham, Dorset BH20 6BB Tel: (0929) 462314

Dr John Eaton, Dept of Botany, University of Liverpool, P.O. Box 147, Liverpool L69 3BX Tel: (051 709) 6022 Ext. 2391/6

8 Sprayers

Sprayers are used to apply liquid formulations of herbicides.

TYPES OF SPRAYER

The two major spraying devices used for aquatic weed control are:

- knapsack – low volume sprayers used by pedestrians or from boats.
- tractor mounted – high volume devices with side arm attachment.

Knapsack (12–25 litres capacity) (*Fig 42*)

These sprays are used for spraying small areas and for spot treatments. They may be:

- hand operated, *ie* sprays an even flow of liquid with each stroke of the pump.
- pressurised – *ie* with a built-in hand pump. Safety valve prevents tank being over-pressurised. One pressurisation is often enough to empty the tank.

Fig 42 Equipment required for knapsack spraying

Tractor mounted (140–1400 litres capacity) (*Fig 30*)

For large areas where access is good. Control valves determine how fast the quantity of liquid is sprayed. Pressure is produced from the tractor power take-off.

APPLICATION RATE

The rate at which the spray is applied depends on the pressure, nozzle size, forward speed and spray width.

For any type of spraying some calibration should be done with each sprayer – this will ensure that the correct dosage is applied to a given area. *Failure to do this will result in high wastage of expensive chemicals plus the danger of grossly over- or under-treating.*

MIXING (for liquids)

- Part fill tank with water (through filter) to prime pump *before* adding chemical. *Never* add concentrated chemical to an empty sprayer tank.
- *Measure* the calculated amount of chemical *do not guess* the amount. Add to the half full tank (through the filter) mixing all the time.
- Wash out used measuring containers with clean water and add to tank.
- Top up tank by keeping the hose below the surface of the mixture to avoid foaming.
- Wash any spilt chemicals off the sprayers and containers plus boots, gloves etc.
- *NB.* When using small knapsack sprayers, herbicides can be mixed in a plastic bucket before adding to tank.

AFTER SPRAYING

- Return unused chemicals to safe storage.
  Label containers to indicate the nature and dilution of chemicals included.
  Dispose of empty containers.
  Dispose of any spray liquid in the tank.
  Dispose of the tank washings.
- Empty sprayer completely and decontaminate with clean water (if possible allow to stand overnight).
- Store sprayer according to manufacturers instructions.

USEFUL PUBLICATIONS (for hand/tractor spraying)

Guidelines for Applying Crop Protection Chemicals. MAFF Booklet 2272 (1983)
The Health and Safety (Agriculture) (Poisonous Substances) Regulations 1975, from HMSO
Guidelines for the Disposal of Unwanted Pesticides and Containers on Farms and Holdings. ADAS Booklet 2198
Storage of Pesticides on Farms. From the Health and Safety Executive
Control of Pesticides Regulations 1986 (SI 1986/1510)

9 Further reading

WEED CONTROL IN FISHERY MANAGEMENT

Fishery Management and Keepering – R Seymour (1970) Charles Knight & Co. Ltd, London
The Management of Angling Waters – A Behrendt (1977) Andre Deutsch

PLANT IDENTIFICATION

British Water Plants – S Haslam, C Sinker, P Wolseley (1975) Field Studies Council
A Handbook of Water Plants – E M Bursche (1971) Warne
A Guide to Identifying British Aquatic Plant Species (1982) Nature Conservancy Council
Aquatic Plants – A Guide to Recognition (1986) ICI Publications
A Beginners Guide to Freshwater Algae – Belcher, Swale (1976) Institute of Terrestrial Ecology

HERBICIDE USE

Guidelines for the use of herbicides on weeds in or near water courses and lakes (1985) MAFF Booklet 2078
Pesticides: Guide to the new controls (1987) MAFF leaflet UL79

GENERAL

Waterways and Wetlands (1981) British Trust for Conservation Volunteers

Index

Algae 16, 85–93
  Blooms 17, 51, 58, 87
  Blue-green 91–94
  Filamentous 17, 67, 88–91
  Unicellular 85–88
Amphibious bistort 126
Arrowhead 121

Black plastic 38–39, 137
Blanket weed see Algae – filamentous
Booms 33
Broad-leaved pondweed 123
Buckets 35, 37, 95, 137
Bulrush see Reedmace or Club-rush
Bur-reed 117, 118
Bypassing 44, 45

Callitriche spp see Starwort
Canadian pondweed 128–129
Carex species see Sedges
'Casoron' see Dichlobenil
Ceratophyllum spp see Hornwort
Chalk streams 96–102
Chemical weed control 62–84, 98, 137
Cladophora spp see Algae – filamentous

'Clarosan' see Terbutryne
Club-rush 118, 119
Common carp 55–58, 91
Common reed 115
Copper sulphate 88, 93
Cott see Algae – filamentous
Crayfish 59–61, 91
Cutting 32, 137
  Cutting boats 34, 138
  Cutting days 101
Curled pondweed 131

Dalapon 118
Deoxygenation 63–64
Depth 42
Dichlobenil 66–67, 75–78
Digging 31
Diquat 66–67, 78–81
Draining 45–46
Duckweed 122

Elodea spp see Canadian pondweed
Enteromorpha spp see Algae – filamentous

Fennel-leaved pondweed 132–133

Glyceria spp see Reed sweet-grass
Glyphosate 67–68, 71–74
Grass carp 48–55, 91, 137

Herbicides see Chemical weed control
Hippuris spp see Mare's tail
Hornwort 128–129
Hydrodicton see Algae – filamentous

Legislation   63
Lemna spp   see Duckweed
Lilies (water)
   White water-lily   124–125
   Yellow water-lily   124–125
Links   32

Mare's-Tail   119–120
'Midstream'   see Diquat
Milfoil   130
Mirror carp   see Common carp
Myriophyllum spp   see Milfoil

Norfolk reed   see Common reed
Nuphar spp   see Lily – yellow
Nutrients   18, 42–44
Nymphaea spp   see Lily – white

Phalaris spp   see Reed canary-grass
Phragmites spp   see Common reed
Polygonum spp   see Amphibious bistort
Potamogeton crispus   see Curled pondweed
Potamogeton natans   see Broad-leaved pondweed
Potamogeton pectinalis   see Fennel-leaved pondweed

Rakes   33
Ranunculus spp   see Water-crowfoot
Reed canary-grass   114
Reedmace   116
Reed sweet-grass   112, 113
'Reglone'   see Diquat
Rhizoclonium spp   see Algae – filamentous
Rivers   94–102
   Slow-flowing   94–96
   Fast-flowing   96–102
River purification boards   145

Rorripa spp   see Water-cress
'Roundup'   see Glyphosate

Sagittaria spp   see Arrowhead
Schoenoplectus spp   see Club-rush
Scythes   32
Sedges   112, 113
Silver carp   58–59
Sparganium spp   see Bur-reed
'Spasor'   see Glyphosate
Spirogyra spp   see Algae – filamentous
Sprayers   147–148
  Knapsack   147
  Tractor-mounted   148
Starwort   127
Stigeoclonium spp   see Algae – filamentous
Straw   91

Terbutryne   66–67, 81–84, 90–91
Trees   39–42
  Comparative costs   137
Typha   see Reedmace

Ulothrix spp   see Algae – filamentous

Vaucheria spp   see Algae – filamentous

Water authorities   57, 63, 141–144
Water-cress   120–121
Water-crowfoot   132–133
Winches   37

Books published by
Fishing News Books Ltd

Free catalogue available on request

Advances in fish science and technology
Aquaculture practices in Taiwan
Aquaculture training manual
Aquatic weed control
Atlantic salmon: its future
Better angling with simple science
British freshwater fishes
Business management in fisheries and aquaculture
Cage aquaculture
Calculations for fishing gear designs
Commercial fishing methods
Control of fish quality
The crayfish
Culture of bivalve molluscs
Design of small fishing vessels
Developments in fisheries research in Scotland
Echo sounding and sonar for fishing
The edible crab and its fishery in British waters
Eel culture
Engineering, economics and fisheries management
European inland water fish: a multilingual catalogue
FAO catalogue of fishing gear designs
FAO catalogue of small scale fishing gear
Fibre ropes for fishing gear
Fish and shellfish farming in coastal waters
Fish catching methods of the world
Fisheries oceanography and ecology
Fisheries of Australia
Fisheries sonar
Fishermen's handbook
Fishery development experiences
Fishing boats and their equipment

Fishing boats of the world 1
Fishing boats of the world 2
Fishing boats of the world 3
The fishing cadet's handbook
Fishing ports and markets
Fishing with light
Freezing and irradiation of fish
Freshwater fisheries management
Glossary of UK fishing gear terms
Handbook of trout and salmon diseases
A history of marine fish culture in Europe and North America
How to make and set nets
Introduction to fishery by-products
The lemon sole
A living from lobsters
Making and managing a trout lake
Managerial effectiveness in fisheries and aquaculture
Marine fisheries ecosystem
Marine pollution and sea life
Marketing in fisheries and aquaculture
Mending of fishing nets
Modern deep sea trawling gear
More Scottish fishing craft and their work
Multilingual dictionary of fish and fish products
Navigation primer for fishermen
Netting materials for fishing gear
Ocean forum
Pair trawling and pair seining
Pelagic and semi-pelagic trawling gear
Penaeid shrimps — their biology and management
Planning of aquaculture development
Refrigeration on fishing vessels
Salmon and trout farming in Norway
Salmon farming handbook
Scallop and queen fisheries in the British Isles
Scallops and the diver-fisherman
Seine fishing
Squid jigging from small boats
Stability and trim of fishing vessels
Study of the sea
Textbook of fish culture
Training fishermen at sea
Trends in fish utilization
Trout farming handbook
Trout farming manual
Tuna fishing with pole and line